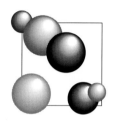

Molecular Biology
in Histopathology

Molecular Medical Science Series

Series Editors

Keith James, University of Edinburgh Medical School, UK
Alan Morris, University of Warwick, UK

 Molecular Medical Science Series

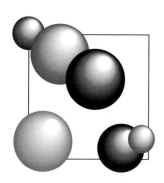

Molecular Biology in Histopathology

Edited by
JOHN CROCKER
East Birmingham (Heartlands) Hospital,
Birmingham, UK

JOHN WILEY & SONS
Chichester · New York · Brisbane · Toronto · Singapore

Other Wiley Editorial Offices

John Wiley & Sons, Inc., 605 Third Avenue,
New York, NY 10158-0012, USA

Jacaranda Wiley Ltd, 33 Park Road, Milton,
Queensland 4064, Australia

John Wiley & Sons (Canada) Ltd, 22 Worcester Road,
Rexdale, Ontario M9W 1L1, Canada

John Wiley & Sons (SEA) Pte Ltd, 37 Jalan Pemimpin #05-04,
Block B, Union Industrial Building, Singapore 2057

Library of Congress Cataloging-in-Publication Data

Molecular biology in histopathology / edited by John Crocker.
        p.   cm. — (Molecular medical science series)
    Includes bibliographical references and index.
    ISBN 0 471 94093 3
    1. Pathology, Molecular.   2. Histology, Pathological.
    I. Crocker, J.   II. Series.
    [DNLM: 1. Cells—pathology.   2. Cells—physiology.   3. Genetic Techniques.
    QZ 50 M71826   1994]
    RB43.7.M6336   1994
    611'.018—dc20
    DNLM/DLC
    for Library of Congress   93–50921
                                                                       CIP

British Library Cataloguing in Publication Data

A catalogue record for this book is available from the British Library

ISBN 0 471 94093 3

Typeset in 11/12pt Palatino by Vision Typesetting, Manchester
Printed and bound in Great Britain by Biddles Ltd, Guildford, Surrey

This book is dedicated to Kate and Stephen, my best friends, who also happen to be my wife and son!

Contents

Contributors

Richard F. Ambinder
Associate Professor of Oncology, John Hopkins Oncology Center, 418 N Bond Street, Baltimore, MD 21231, USA

Mark J. Arends
Lecturer in Pathology, CRC Laboratories, Department of Pathology, University of Edinburgh Medical School, Teviot Place, Edinburgh EH8 9AG, UK

Richard S. Camplejohn
Senior Lecturer, Richard Dimbleby Department of Cancer Research, UMDS, St Thomas' Hospital, London SE1 7EH, UK

John Crocker
Consultant Histopatholgist, Histopathology Department, Birmingham (Heartlands) Hospital, Bordesley Green East, Birmingham B9 5ST, UK and Visiting Senior Clinical Lecturer, University of Warwick, UK

Massimo Derenzini
Centro di Patologia Cellulare, Dipartimento di Patalogia Sperimentale, Universita di Bologna, Via San Giacomo 14, 40126 Bologna, Italy

Kenneth A. Fleming
Clinical Reader in Pathology, Nuffield Department of Pathology and Bacteriology, Level 4, Academic Block, John Radcliffe Hospital, Oxford OX3 9DU, UK

David J. Harrison
Senior Lecturer, CRC Laboratories, Department of Pathology, University of Edinburgh Medical School, Teviot Place, Edinburgh EH8 9AG, UK

Simon G. Long
Clinical Research Fellow, Department of Haematology, Birmingham (Heartlands) Hospital, Bordesley Green East, Birmingham B9 5SS, UK

Adrienne L. Morey
*Clinical Lecturer in Pathology, Nuffield Department of Pathology and
Bacteriology, Level 4, Academic Block, John Radcliffe Hospital, Oxford OX3
9DU, UK*

Paul G. Murray
*Senior Lecturer in Biomedical Sciences, University of Wolverhampton, Molineux
Street, Wolverhampton WV1 1SB, UK*

Dominique Ploton
*Associate Professor of Cell Biology, Faculty of Medicine and Unite de
Recherche, INSERM 314, Reims, France*

Jonathan J. Waters
*Principal Clinical Cytogeneticist, Cytogenetics Laboratory, West Midlands
Regional Genetics Service, Birmingham (Heartlands) Hospital, Bordesley Green
East, Birmingham B9 5SS, UK*

Eric P.H. Yap
*Clinical Research Fellow, Nuffield Department of Pathology and Bacteriology,
Level 4, Academic Block, John Radcliffe Hospital, Oxford OX3 9DU, UK*

Lawrence S. Young
*Senior Lecturer, Cancer Research Campaign Laboratories, Department of Cancer
Studies, University of Birmingham Medical School, Birmingham B15 2TJ, UK*

Preface

The past 20 years have witnessed numerous changes in the practice of histopathology, with many powerful techniques, such as immunohistochemistry and image analysis, aiding the accuracy and objectivity of diagnosis and research. However, perhaps the greater revolution, occurring in the past half decade, of the application of molecular methods, will be even more fruitful. Molecules related to, for example, hormones, immunoglobulins, infectious agents or chromosomes can be identified by means of gene probes. Furthermore, the molecular basis of cell replication has become more clearly understood, assisting in tumour prognosis. What, then, are these new methods? As the title of this book implies, the techniques included in it are those that can be performed on histological material, although not necessarily involving microscopic examination. The reader should not be led into the belief that 'molecular' must always imply 'DNA' and, as we can see in at least two chapters herein, 'molecular' should have a wider, more appropriate meaning.

The purpose of this series of volumes is to supply a guide to those just qualified and undertaking research or to those who have taken degrees some years in the past and who wish to glean new information rapidly. Thus, this is not a molecular 'recipe book'; such exist elsehwere.

In the first chapter, Mr Murray and Professor Ambinder have given an account of the methods available for the demonstration of infectious agents *in situ* in histological material. The applications of these techniques are also outlined. Chapters 2 and 4 by Drs Fleming, Morey and Yap, and by Drs Waters and Long, then describe these methodologies and others as applied to the examination of malignant tissues and chromosomes in histological material.

Chapters 3 and 5 give details of other methodologies, both of which are not histological (although one may become so). These techniques do, however, employ histological material, even of archival, paraffin wax-embedded type. Thus, Dr Young describes the value of the polymerase chain reaction in histopathology; indeed, this is already being adapted for use as an *in situ* method. Dr Camplejohn then describes the techniques of DNA flow cytometry and their applications. One of the latter is that of the assessment of cell proliferative status. Leading on naturally from this, in Chapter 6 I have given an account of the molecular basis of the cell cycle and of some of the antibody probes which can be applied to visualize some of the components of the cell cycle. Also highly related to cell proliferation is the activity of the interphase nucleolar organizer region. The full significance of this structure is not yet fully understood but in the subsequent chapter, Professors Derenzini and Ploton give an account of the morphological and molecular corollaries of the nucleolar organizer regions.

Just as we are realizing and understanding the importance of cell proliferation in disease, so we are appreciating that cell death is also central to many physiological and pathological conditions. Accordingly, in Chapter 8, Drs Arends and Harrison tell us of the molecular basis of 'programmed cell death' or apoptosis in health and disease.

Thus, this volume gives an introduction to the currently available molecular techniques in histology and an account of the molecular basis of certain phenomena of importance in everyday histopathology.

John Crocker
Birmingham
1994

Acknowledgements

I am most grateful to Lesley Winchester, Lucy Jepson and Verity Waite of John Wiley & Sons for their constant help over the gestation of this book, and to Dr Alan Morris of Warwick University for suggesting it.

Of course, I am endebted to the contributors to this volume for their excellent chapters.

I also thank Miss Jane Oates for her helpful discussion of many topics included here.

Finally, I thank my wife, Kate and son, Stephen, for their patience during many long hours of writing and editing.

1 *In Situ* Hybridization in Relation to Infectious Agents

P.G. MURRAY and R.F. AMBINDER

Recent advances in molecular biology have provided the tools to revolutionize diagnostic histopathology. In particular *in situ* hybridization (ISH) is a technique which is being increasingly applied in the diagnostic arena. ISH enables the precise identification and localization of a specific nucleic acid sequence within cells or tissue sections and this chapter will examine the basic principles of the method and give an overview of technical considerations. Applications of the procedure to the detection of infectious agents will then be considered with particular emphasis on the detection of viral sequences. Other chapters (2 and 4) consider the application of ISH to the detection of oncogenes and oncogene products and to the analysis of chromosome abnormalities.

BASIC PRINCIPLES OF ISH

ISH involves the detection of specific sequences of DNA or RNA within tissue sections or cell preparations by means of a labelled complementary nucleic acid sequence or probe. Under appropriate conditions the probe will hydrogen bond (hybridize) to the target DNA or RNA. In this way stable hybrids will be formed which can be visualized by a variety of means depending on the nature of the label used (Fig. 1.1). If the target or probe nucleic acid is double-stranded it will have to be made single-stranded before hybridization can take place. The process of making double-stranded nucleic acid single-stranded is known as denaturation and is usually achieved by the application of heat. The temperature at which denaturation takes place is referred to as the melting temperature. Successful ISH depends upon tissue- and probe-related factors, hybridization conditions and the detection system.

Molecular Biology in Histopathology. Edited by J. Crocker
© 1994 by John Wiley & Sons Ltd

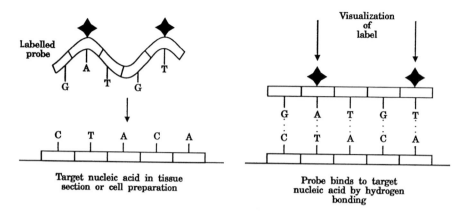

Fig. 1.1. Labelled nucleic acid sequence (probe) hybridizes through complementary base pairing to a target sequence in the tissue section or cell preparation. Hybridization is detected by visualizing the label

TISSUE-RELATED FACTORS

Tissue preparation

In routine histopathology, the emphasis is usually on morphological analysis employing formaldehyde fixation and paraffin wax embedding to generate tissue sections. Both DNA and RNA can be detected readily by ISH in tissue processed in this way. However, there are a number of points worthy of consideration.

Delay in fixation should be avoided since enzymes such as endogenous ribonucleases may degrade the target nucleic acids, resulting in a poor hybridization signal. Whereas DNA is less susceptible to degradation, RNA is readily digested by RNase and prompt fixation is therefore necessary to inactivate any endogenous RNase present.

Tissue fixed in cross-linking fixatives such as formaldehyde or paraformaldehyde will normally require a permeabilization step prior to hybridization. This allows the probe access to tissue nucleic acids and is generally achieved by the use of proteolytic enzymes such as pronase, although in certain procedures hydrochloric acid or triton X-100 may be employed as additional permeabilization steps. Lengthy fixation should also be avoided as this has been shown to reduce hybridization efficiency.

Paraffin wax sections should be cut directly onto slides coated with an adhesive since sections have a tendency to float off during the hybridization step. In our experience aminopropylethoxysilane (APES) has proved most effective, although other section adhesives may be suitable.

Cryostat sections or cell smears are also suitable for ISH and often yield hybridization signals of greater intensity than paraffin material. They may be necessary if the process is to be combined with the immunohistochemical demonstration of antigens that are poorly preserved in formalin.

Non-specific binding

To block non-specific binding of the probe to components of the tissue section a blocking solution such as Denhardt's is often included in pre-hybridization and/or hybridization steps. Denhardt's solution contains nucleic acid (salmon sperm DNA or yeast RNA), another polyanion (polyvinyl pyrrolidone), detergent (sodium dodecyl sulphate) and albumin. A variety of other blocking solutions can also be used.

Nature of the target sequence

Both DNA and RNA are suitable targets for ISH. In terms of the analysis of viral infections the choice of target will depend in part upon the nature of the infection itself. For example, viral DNA may be the target of choice in latent infections since there may be little or no transcription. Similarly, with lytic infection, DNA may be a good target because of the high copy number. RNA is most useful to target, when, as with latent Epstein–Barr virus (EBV) infection, the DNA target may be in limited copy number but the RNA copy number is high. The detection of viral RNA may also provide useful information about the pattern of viral gene expression (Fleming, 1992).

PROBE-RELATED FACTORS

Types of probe

The main types of probe currently used for ISH are cloned RNA and DNA probes (either single-stranded or double-stranded) and oligonucleotide probes (Zeller and Rogers, 1991). RNA probes are produced by *in vitro* transcription of DNA sequences cloned into expression vectors such as plasmids. Expression vectors with the cDNA running in opposite directions permits the production of antisense transcripts which act as the probe and sense transcripts which can act as negative controls. Transcription in the presence of labelled nucleotides will produce a labelled probe. RNA probes have less non-specific binding than do DNA probes and they may be shortened by alkaline hydrolysis to facilitate their access to tissue. In addition, post-hybridization digestion in RNase may be employed to remove all non-hybridized single-stranded RNA probe and thereby reduce background signal.

DNA probes are generally considered to be more robust than RNA probes, mainly because they are not subject to digestion by ubiquitous RNases. Double-stranded DNA probes have the advantage that the strands can form networks, thus amplifying the signal. Single-stranded probes may be prepared if single-stranded phage (e.g. M13) or phagemid (plasmid with phage origin of replication, e.g. BlueScribe M13) is employed. The preparation of single-stranded DNA is more difficult than generating labelled single-stranded RNA and is therefore not widely used. Single-stranded probes have several advantages. In general they will hybridize more efficiently than double-stranded probes and unlike double-stranded probes will not self-anneal.

Oligonucleotide probes are oligomers of DNA nucleotides and are generally produced on automated oligonucleotide synthesizers. In order to synthesize these probes one must first know the sequence of the target. This can be obtained from computerized banks of nucleic acid sequences or may be inferred from the amino acid sequence of the protein product. Careful selection of the oligonucleotide sequence is necessary to reduce the risk of hybridization to irrelevant genomic sequences.

Ideally, the oligonucleotide sequence should be rich in guanine and cytosine nucleotides since these will form a more stable hybrid (these two bases are associated with greater hydrogen bonding than adenine and thymine). The size of oligonucleotide probes is limited to the size that can be synthesized efficiently. The yield of full-length product decreases as the number of bases increases. Probes between 18 and 50 base pairs (bp) in length are synthesized readily and give adequate signals for moderately abundant target sequences. Target sequences are often accessible to these short probes even without treatment of specimens with permeabilizing agents.

Probe labelling

Labels consist of either radioactively or non-radioactively labelled nucleotides incorporated into the nucleic acid probe. Most early ISH was performed using radioactive probes which were then detected by autoradiography. A number of radioisotopes are available for labelling probes, and each has its own benefits and drawbacks. ^{32}P, for example, has a half-life of only 14 days, and so probes incorporating this label are not stable for long periods. Exposure time are short but resolution is poor, and ^{32}P does not provide accurate cellular localization. ^{125}I has a half-life of 60 days, requires only short exposures and gives high resolution; however, it is often associated with high background and requires special precautions for handling. ^{35}S has a half-life of 87 days, requires moderate exposure times, provides acceptable resolution, and is often the isotope of choice. ^{3}H has a half-life of 12 years, requires long exposure and gives excellent resolution.

When using radioactive probes, specimens must often be duplicated in order to obtain the optimum exposure. In some cases, the interpretation of the final preparation may be difficult since photographic silver grains may obscure cell morphology. In general, however, the method affords high sensitivity which is often enhanced if long autoradiographic exposure times are employed.

The disadvantages associated with the use of radioactive probes have often precluded their use in routine histopathology. In this situation non-isotopic probes are more desirable. Non-isotopic probes may be labelled with peroxidase, biotin or digoxigenin. Unlike biotin, digoxigenin is not a natural constituent of mammalian tissue but occurs exclusively in *Digitalis* spp. Detection using these non-radioactive labels is somewhat less sensitive than autoradiography but it is now possible to detect single copy sequences by means of these systems.

As described previously, RNA probes are labelled conveniently during the *in vitro* transcription stage. DNA probes may be labelled in one of two ways. In the nick translation method, the enzyme DNAse I produces small breaks or 'nicks' in a DNA strand. Then the $3'-5'$ exonuclease activity of the enzyme DNA polymerase I excises small groups of nucleotides adjacent to the nicks and fills in the gaps with labelled nucleotides. In the random primer extension technique random oligonucleotides (hexamers) are annealed to a single-stranded probe. The Klenow fragment of DNA polymerase extends the primer incorporating labelled nucleotide bases. The benefits of this method are the high specific activity of the probe and the ability to label very short DNA fragments.

Oligonucleotide probes are labelled at either or both ends. T4 polynucleo-tide kinase is used to label the $5'$ end and terminal deoxynucleotide transferase to label the $3'$ end. End-labelled probes are less sensitive than larger body-labelled probes which incorporate more labelled nucleotides per probe molecule. Because of their greater sensitivity, the larger cloned probes are often required for the detection of low copy number target sequences.

HYBRIDIZATION CONDITIONS

Hybridization occurs when complementary base pairs form between the labelled probe and the target sequence. This is favoured by conditions that facilitate hydrogen bonding (high salt concentrations and low temperatures, i.e. usually 25 °C less than the temperature at which the hybrids dissociate). Although specific hybrid formation will readily occur under such conditions, imperfectly matched hybrids will also be formed. Since the hydrogen bonds between mismatched hybrids are weaker than hybrids in which there is 100% base pair complementation, their formation is inhibited by conditions less favourable to hydrogen bonding. The optimal conditions for hybridization will therefore be a compromise between the desire for specific hybrid formation and the elimination of mismatched hybrids. Conditions which favour hydrogen bond formation and therefore also mismatched hybrid formation are said to be of low stringency. In contrast, those conditions which are less favourable for hydrogen bonding and in which mismatched hybrid formation is reduced are said to be of high stringency. Stringency may also be increased by the use of several post-hybridization high-temperature washes in salt solutions of decreasing strength. Formamide is often included in hybridization solutions and lowers the melting temperature of nucleic acid hybrids, thus allowing higher stringency at the lower temperature range, which is beneficial for retention of tissue on the slide and preservation of morphology. The exact conditions for hybridization will depend upon the nature of the hybrids formed. For example, RNA–RNA hybrids have a higher melting temperature than DNA–DNA hybrids and therefore stringent hybridization for RNA–RNA is achieved at temperatures around 10–15 °C higher than those required for DNA–DNA.

DETECTION SYSTEMS

Radiolabelled probes are detected by autoradiography. Despite its disadvantages, this method may have some useful applications, notably in the quantitation of signal by grain counting.

Detection systems for peroxidase, biotin and digoxigenin are readily available and easy to use, generating a signal within hours as opposed to days for autoradiography. Peroxidase-labelled probes are readily detected by the diaminobenzidine reaction, which generates a brown reaction product, whereas biotin may be detected by the use of avidin–biotin systems or anti-biotin antibodies conjugated to either peroxidase or alkaline phosphatase. In the case of digoxigenin, detection is usually based upon the use of anti-digoxigenin antibodies conjugated to alkaline phosphatase.

CONTROLS

In order to validate the results obtained by ISH, appropriate controls should be employed. These normally include positive controls in the form of material known to contain the target sequence and suitable negative controls, including the use of a 'no probe' control. The 'no probe' control will identify both non-specific binding of reagents used to detect the label, and the presence of any endogenous enzyme activity in cases where an enzyme has been used as the label or in the detection system. Antisense probes are used to detect RNA, and sense probes serve as negative controls. In order to evaluate the integrity of nucleic acid in a tissue specimen and its accessibility to probe, a positive control probe is commonly used. This is particularly important when the target sequences are RNA, which are very susceptible to degradation. Poly-d(T) probes, which detect polyadenylated mRNA transcripts, are often used for this purpose when the hybridization target is mRNA. Similarly, when detecting polymerase III viral transcripts, we have found it useful to probe for cellular RNA polymerase III transcripts, notably the cellular U6 RNAs.

ISH FOR THE DETECTION OF INFECTIOUS AGENTS IN GENERAL

ISH has many applications in the detection and analysis of viral infections and these are discussed in detail later below. However, when considering the importance of the technique in infectious disease it is often forgotten that ISH can be used to detect many non-viral targets, including bacteria, protozoa and even some helminths.

The direct detection of infectious agents either in tissue sections or in other clinical samples is an attractive diagnostic option as it does not rely on the *in vitro* growth of the organism which for some organisms (e.g. *Mycobacterium tuberculosis*) can take up to several weeks, or in the case of others may even be impossible (e.g. *M. leprae*). Conventional detection of mycobacterial infections in tissue sections is also limited to the determination of 'acid-fast bacilli' using techniques such as the Ziehl–Neelson method. In addition, routine microbiological assessment of mycobacteria suffers from the major drawback that a species-specific diagnosis is extremely time-consuming and in several cases even impossible. ISH offers a more acceptable alternative and has been successfully employed for the rapid detection and differentiation of mycobacterial species in both fresh frozen and routinely processed paraffin wax sections of infected tissues. There are other situations when it is desirable to detect a pathogen directly. For example, as a result of the severe underlying immune dysfunction many opportunistic infections in patients with AIDS cannot be reliably diagnosed by serology. In addition serological responses to an infectious agent may not always be present at all stages of infection.

ISH can also be used as an adjunct to the morphological identification of many infectious agents. Often the microscopical detection of an organism may be time-consuming and accurate identification may require considerable experience, particularly when searching for the organism in tissue smears or faecal preparations. ISH offers the advantage of recognisable organism morphology in combination with hybridization signal and in some situations has proved to be more sensitive than morphology alone, notably in the detection of *Plasmodium falciparum* in blood films. However, ISH has been shown to be less sensitive than conventional Papanicolaou staining for the detection of *Chlamydia trachomatis* in cervical smears.

Through the development of probes specific for virulence factors it is now possible to distinguish pathogenic and non-pathogenic organisms. This approach has been used to detect enteropathic *Escherichia coli* against the extensive background of non-pathogenic, so called commensal, strains. Similarly virulent species of *Yersinia enterocolitica* and *Campylobacter jejuni* have been identified in this way.

Much of the early ISH was performed using probes directed to the DNA of the target organism. More recently, however, rRNA has proved a more suitable target in some cases, and since it is often present in high copy number this approach is more sensitive than the detection of DNA, such that in many cases large probes are not required and labelled oligonucleotides usually suffice. Most detection systems currently in use are based upon digoxigenin or biotin as previously described. However, direct and more rapid analysis is possible if fluorescent labelled probes are employed.

ISH may also allow novel approaches to the study of infections. Thus, ISH has shown that mRNA for the chicken homologue of the human prion protein (PrPC, scrapie isoform of prion protein) is widely distributed in cholinergic and non-cholinergic neurones throughout the adult central nervous system, suggesting that the protein has a more widespread role other than the regulation of the acetylcholine receptor. In *Trypanosoma cruzi* infections ISH has been used to localize insertion of *T. cruzi* DNA to specific chromosomes.

APPLICATIONS OF ISH TO THE DETECTION AND ANALYSIS OF VIRAL AGENTS

GENERAL CONSIDERATIONS

In the past, the detection of viruses in tissue sections or cell preparations has relied upon either morphological identification of the characteristic cellular changes associated with infection or on the use of tinctorial procedures such as the phloxine–tartrazine method. More selective methods are available such as Shikata's orcein method, which detects the hepatitis B surface antigen found in hepatitis B virus infections. Many of these methods rely upon certain histochemical properties of viral products; thus Shikata's method is considered to be dependent upon the presence of high concentrations of cysteine within the hepatitis B surface antigen. The sensitivity of these methods is generally low, and in most cases latent infections cannot be demonstrated. Immunohistochemistry has proven useful in many situations but is limited, particularly in latent viral infection where expression may be confined to a small number of viral genes. In EBV-associated Burkitt's lymphoma, for example, only one viral protein, the EB nuclear antigen-1 (EBNA-1), is expressed in infected tumour cells, and reliable antibodies are not available for EBNA-1 detection in paraffin wax sections. ISH offers the opportunity directly to detect the viral genome within a cell or tissue and to characterize the pattern of viral gene expression.

ISH may offer certain advantages when compared to the traditional methods of virus detection, which include culture, electron microscopy or serology. For example, the human papillomaviruses (HPVs) cannot be isolated from clinical specimens by culture methods, and electron microscopy cannot detect HPV in latent infection. Using immunohistochemistry, HPV capsid antigens can be detected but only in productive infections. In contrast, ISH permits the detection of HPV in clinical specimens in both productive and latent infections. Furthermore, the use of subtype-specific probes has been important in establishing the association of certain HPV subtypes with specific malignancies. For example, HPV subtypes 16 and 18 are commonly detected in cervical cancer, whereas HPV subtypes 5 and 8 have been shown to be associated with some cases of squamous carcinoma of the skin.

ISH may also provide a rapid method for the detection of viruses in clinical specimens. For example, it has been used for the rapid diagnosis of cytomegalovirus (CMV) infections in liver grafts following transplantation (Naoumov *et al.*, 1988). ISH has also been used to search for potential viral involvement in diseases of unknown aetiology. CMV, for example, has been localized to the islets of Langerhans in some cases on non-insulin-dependent diabetes mellitus, suggesting a potential role for the virus in the pathogenesis of this disorder (Lohr and Oldstone, 1990).

In theory, the range of viral nucleic acid sequences that can be detected by ISH is only limited by the availability of clones or sequence information. Many important human viruses, including, for example, the human immunodeficiency viruses (HIV) 1 and 2, EBV and HPV, have now been completely sequenced and this information is readily available. A number of viral genes have some homology to cellular genes. For example, an EBV gene has regions of marked homology with a cellular interleukin gene. Probes should be carefully selected to avoid regions of shared homology which would otherwise result in hybrid formation between probe and inappropriate cellular sequences.

The polymerase chain reaction (PCR) provides a sensitive alternative for the detection of viruses in clinical specimens (Chapter 3). It has been estimated that PCR can detect one copy of sequence in 10 000 cells. However, the detection of viral infection by PCR analysis does not provide any information as to the cell type harbouring the virus, or as to what fraction of cells is infected. ISH enables the precise cellular location and extent of viral infection to be determined.

RECENT DEVELOPMENTS

There have been several recent reports describing a technique which combines PCR with ISH and permits the localization of specific amplified DNA segments within isolated cells and tissue sections. *In situ* PCR has been reported to be more sensitive than ISH alone and may be useful for the detection of low-copy-number target sequences. It has been used, for example, to detect low-copy HIV-1 in peripheral blood mononuclear cells when HIV infection could not be detected by ISH (Bagasra *et al.*, 1992).

Another important recent development has been the combination of immunohistochemistry and ISH. Thus, we have been able to identify specific cell types infected by EBV by ISH followed by visualization of cell-specific antigens by immunohistochemistry in a variety of situations, some of which are discussed in more detail later in the chapter. Other examples of this approach include the demonstration of the human parvovirus B19, not only in erythroid cells as expected, but also in mononuclear phagocytes and myocardial cells (Porter *et al.*, 1990). Similarly, it is possible to examine virally infected cells for the presence of a specific host cell protein whose expression might be affected by the presence of the virus.

Simultaneous demonstration of two virus types by two differently labelled probes is also possible. Thus, HPV and herpes simplex virus genome have been co-localized in this way (Mullnink *et al.*, 1989). Similarly, double labelling of virus and individual chromosomes by ISH has enabled the localization of viral integration sites to specific chromosomes. As with all double labelling approaches, care must be taken to ensure that the detection systems do not interfere with each other.

Ultrastructural ISH has met with some success using colloidal gold, and should allow the localization of replication and viral gene expression within the cell at different stages of infection. The confocal laser microscope is another development which may provide insight into viral pathogenesis, particularly in the area of localization of viral integration.

ILLUSTRATION OF THE APPLICATION OF ISH TECHNIQUES TO CHARACTERIZATION OF EBV IN CLINICAL TISSUES

ISH is particularly well suited to answering questions about viral localization within a histologically complex tissue. For example, Hodgkin's tumours are a polymorphic mixture of lymphocytes, neutrophils, eosinophils, monocytes and malignant cells known as Reed–Sternberg and Hodgkin cells. The Reed–Sternberg cells often consistute less than 1% of cells in tumour masses. Detection of EBV DNA by Southern blot hybridization has not indicated whether the virus is present in benign infiltrating B lymphocytes, tumour cells, or both. If EBV was only associated with lymphocytes and not with the malignant cells, then one might be inclined to dismiss its presence as an epiphenomenon. On the other hand, a consistent association with Reed–Sternberg cells would suggest a role in pathogenesis.

An *in situ* technique allows the association of virus with a particular cell population. However, the application of *in situ* detection methods to pathological specimens for the detection of latent EBV is limited by the highly restricted expression of the EBV genome in malignancy. As noted above, only one EBV protein, EBNA-1, is known to be expressed in all EBV-associated tumours, and methods to detect this antigen have not been reliable when applied to formalin-fixed tissues. Study of EBV-immortalized B lymphoblastoid cell lines suggest an ideal target for ISH. These cell lines express very short viral RNA transcripts (165 and 169 nucleotides) in abundance (up to 10^7 copies per cell). These EB early RNA (EBER-1 and EBER-2) transcripts do not code for protein, and in contrast to messenger RNAs, which are transcribed by RNA polymerase II, the EBERs are transcribed by RNA polymerase III. The function

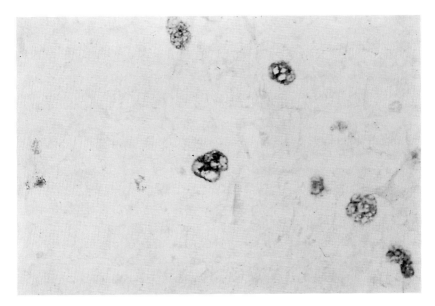

Fig. 1.2. *In situ* hybridization of a Hodgkin's disease biopsy specimen. A digoxigenin-labelled EBER antisense probe demonstrates the presence of latent EBV infection in the Reed–Sternberg cells

of the EBERs is uncertain, but their abundance makes them useful as targets (Ambinder *et al.*, 1993).

In contrast to more commonly targeted mRNAs, the EBERs have a distinctive nuclear localization. This characteristic of the hybridization is frequently useful in distinguishing artefactual signal from true hybridization signal but can only be appreciated with high-resolution detection systems such as ^{3}H or non-isotopic systems. Figure 1.2 shows EBER hybridization in Hodgkin's disease. The hybridization signal is clumped round the nuclear rim and nucleoli. Only Reed–Sternberg cells and their variants are positive.

Generally single-stranded RNA probes ('riboprobes') synthesized *in vitro* from a plasmid template are used. The plasmid template used can be constructed by cloning the product of PCR amplification of EBER-1 sequence from viral DNA into a recombinant plasmid vector between two bacterio-phage promoters (T7 and T3) (Fig. 1.3). The recombinant plasmid is linearized with restriction enzyme and transcribed *in vitro* with T7 polymerase or T3 polymerase in the presence of labelled nucleotides to generate antisense or sense probes. The antisense probe will hybridize with the viral sense transcript and identify the presence of viral RNA. The sense probe does not specifically hybridize to any viral or cellular transcript. Signal associated with the sense probe indicates a flawed assay. Possible problems include specimen preparation and hybridization conditions which allow hybridization with viral DNA, non-specific hybridization with cellular RNA, or may reflect non-specific

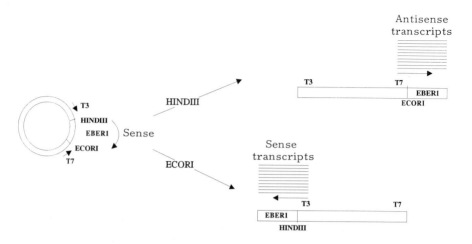

Fig. 1.3. A plasmid for *in vitro* transcription to generate sense and antisense EBER riboprobes

stickiness for probe or reagents used in detection of probe hybridization. Labelled nucleotides used with this system include ^{35}S, ^{3}H, digoxigenin and biotin labels. An alternative to the use of *in vitro*-generated RNA transcripts is the use of automatically synthesized labelled oligonucleotides.

EBER hybridization signal is readily detected in routinely prepared formalin-fixed paraffin-embedded sections. Perhaps because of the abundance of the target, very short hybridization times suffice (Barletta *et al.*, 1993). The entire procedure, including hybridization, antibody detection and colour development, can be carried out in 3 h.

If the EBERs are not detected in a specimen, what evidence is there that the RNA is intact and accessible to probe? In other hybridization systems, as noted above, the integrity and preservation of target RNAs in tissues have been assessed by hybridization to mRNAs associated with 'housekeeping' genes such as actin. However, these mRNAs are rare by comparison with the EBERs and have a different subcellular localization. In order to assess the validity of negative hybridization results, a plasmid library can be constructed for *in vitro* transcription to generate riboprobes to the U6 cellular gene. U6 is similar to the EBERs in subcellular localization, stem–loop structure and transcription by RNA polymerase III. Despite these similarities, the U6 antisense probe does not lead to any cross-hybridization with the EBERs. U6 detection indicates that abundant nuclear polymerase III transcripts are preserved and accessible to probe. A concomitant failure to detect EBERs suggests the absence of EBV latency infection.

ISH not only enables detection of virus but also characterization of patterns of gene expression. EBV lytic infection, because it is associated with many copies of the viral genome, may be readily detected by ISH procedures

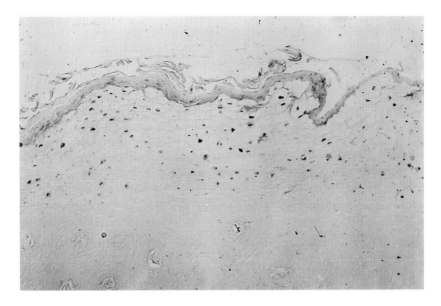

Fig. 1.4. *In situ* hybridization of an oral hair leukoplakia biopsy specimen. A digoxigenin-labelled *Not1* antisense probe demonstrates the presence of lytic EBV infection in the spinous layer of the tongue epithelium

targeting DNA. However, this procedure does not specifically distinguish latent from lytic infection; this is a distinction that may be important since lytic infection is inhibited by antiviral agents such as acyclovir. In order specifically to detect lytic infection, riboprobes have been developed for mRNAs expressed only during lytic infection (Ryon *et al.*, 1993). The *Not1* mRNA has two virtues that facilitate its detection. Firstly, this transcript is among the most abundantly transcribed RNAs early in lytic infection, although the level of RNA expression does not compare with that of the EBERs. Secondly, the *Not1* transcript includes 12 copies of a 125 bp repeated sequence. Repeated sequences are recognized in many diverse systems in biology and make attractive targets for hybridization. This probe generates a strong signal in oral hairy leukoplakia, a lesion associated with intense lytic EBV infection that is reponsive to acyclovir therapy (Fig. 1.4).

Another EBV lytic transcript is of interest in its own right. This transcript codes for a secreted protein sequentially and functionally homologous to the interleukin 10 (IL-10) cytokine. This cytokine has pleiotropic effects on immune function, including inhibition of IL-2 and interferon-γ secretion. In order to determine whether this viral mRNA was being synthesized in pathological specimens, a probe has been prepared that targets only the unique leader sequence of the transcript of the viral gene. In contrast with the *Not1* probe, the viral IL-10 probe is not sensitive in detecting the presence of lytic EBV infection because the target transcript is neither abundant nor repetitive.

However, the viral IL-10 probe does allow unequivocal detection and localization of the viral transcript in clinical specimens.

In summary, ISH is a technique that allows the cellular localization of infection within a tissue specimen, even when the tissue is a complex mixture of cell types. The biology of a particular system studied should guide the choice of target. Targeting sequences with high copy number by virtue of viral replication, viral expression or repeats will facilitate detection. Care must be taken to ensure that the probe used is specific for the sequences targeted and that nucleic acids in the specimen are neither degraded nor inaccessible to probe.

REFERENCES

Ambinder RF, Browning PJ, Lorenzana I et al. (1993) Epstein–Barr virus and childhood Hodgkin's disease in Honduras and the United States. *Blood*, **81**, 462–467.

Bagasra O, Hauptman SP, Lischer HW, Sachs M and Pomerantz RJ (1992) Detection of human immunodeficiency virus type-1 provirus in mononuclear cells by *in situ* polymerase chain reaction. *N. Engl. J. Med.*, **326**, 1385–1391.

Barletta JM, Kingma DW, Ling Y et al. (1993) Rapid *in situ* hybridization for the diagnosis of latent Epstein–Barr virus infection *Mol. Cell. Probes*, **7**, 105–109.

Lohr JM and Oldstone MBA (1990) Detection of cytomegalovirus nucleic acid sequences in pancreas in type 2 diabetes. *Lancet*, **336**, 644–648.

Mullnink H, Walboomers JMM, Raap AK and Meyer CJ (1989) Two colour DNA *in situ* hybridization for the detection of two viral genomes using non-radioactive probes. *Histochemistry*, **91**, 195–198.

Naoumov NV, Alexander GJM, O'Grady JG et al. (1988) Rapid detection of cytomegalovirus infection by *in situ* hybridization in liver grafts. *Lancet*, **i**, 1361–1364.

Porter HJ, Heryet A, Quantrill A and Fleming KA (1990) Combined non-isotopic *in situ* hybridization and immunohistochemistry on routine paraffin wax embedded tissue: identification of cell type infected by human parvovirus and demonstration of cytomegalovirus DNA and antigen in renal infection. *J. Clin. Pathol.*, **43**, 129–132.

Ryon J, Hayward SD, MacMahon EME et al. (1993) *In situ* detection of lytic Epstein–Barr virus infection: expression of the Not1 early gene and vIL-10 late gene in clinical specimens. *J. Infect. Dis.*, **168**, 345–351.

Zeller R and Rogers M (1991) In situ hybridization to cellular RNA. In: Ausubel FM, Brent R, Kingston RE et al. (eds), *Current Protocols in Molecular Biology*, pp. 1403–1468. New York: Greene Publishing Associates and Wiley–Interscience.

FURTHER READING

Fleming KA (1992) Editorial: analysis of viral pathogenesis by *in situ* hybridization. *J. Pathol.*, **166**, 95–96.

Warford A and Lauder I (1991) *In situ* hybridization in perspective. *J. Clin. Pathol.*, **44**, 177–181.

2 Genetic Analysis of Malignancy in Histopathological Material Using Filter and *in Situ* Hybridization

K.A. FLEMING, A.L. MOREY and E.P.H. YAP

Over the last 10 years, technical developments in molecular biology have allowed the application of its investigative power to histopathological material. The two advances of particular relevance are filter hybridization and *in situ* hybridization (ISH). These are of considerable importance since probably the largest resource of human tissue (both normal and diseased) in the world is generated and stored as archival material in histopathology departments. Although most use to date has been in the detection of viruses, many other areas of interest have also been investigated, including chromosomal assignment of gens, studies of intra-nuclear chromosomal organization and studies on gene expression in a variety of sites and circumstances. The investigation of malignancy has been a particularly important area and will be discussed in this chapter.

SPECIFIC TECHNIQUES

FILTER HYBRIDIZATION

In this form of nucleic acid hybridization, target nucleic acid (DNA or RNA) is extracted from cells or tissues, and then immobilized on filters made of nitrocellulose or nylon. This is done either directly as 'dot blots' or after restriction enzyme digestion, size fractionation by gel electrophoresis and subsequent transfer onto the filter (Fig. 2.1). Following this, the bound nucleic acid is hybridized to a labelled probe and after washing the bound probe is detected. Data on labelled probes are given in Tables 2.1, 2.2 and 2.3.

Dot blot Hybridization

Where size fractionation of nucleic acid is not required, for instance when only quantification of the amount of target nucleic acid is required, nucleic acid may

Molecular Biology in Histopathology. Edited by J. Crocker
© 1994 by John Wiley & Sons Ltd

16

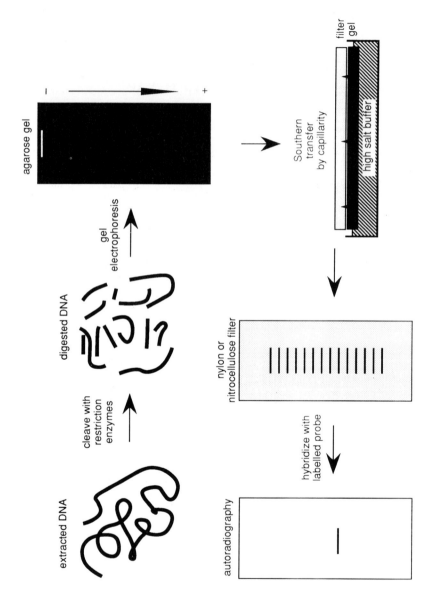

Fig. 2.1. Schematic diagram of Southern blot hybridization. See text for details

Table 2.1. Some characteristics of labelled probes

	Uses	Advantages/disadvantages	
Radioactive labels			
^{32}P	All filter hybridizations (Northern, Southern, dot blot)	Sensitive and relatively fast	Poor resolution for ISH
^{33}P	Filter and ISH	Intermediate between ^{32}P and ^{35}S	
^{3}H	ISH	Good resolution	Long exposure times
^{35}S	ISH	Relatively fast and reasonable resolution	High background
^{125}I	Liquid hybridization and ISH	Fast	Poor resolution Short half-life
Commonly used non-radioactive labels			
Biotin	Filter and ISH	Sensitive	Background due to endogenous biotin in tissues
Digoxigenin	Filter and ISH	Low background, reasonable sensitivity	

Table 2.2. Probe types

Probe	Advantages/disadvantages
Plasmid, phage and cosmid (dsDNA)	Widely available, stable, with large insert capacity, but vector sequences can cross-react with genomic DNA
Plasmid with RNA promoters (riboprobe)	Single-stranded probe (RNA). Sensitive, sense/antisense configuration, but labile
Oligonucleotide (ssDNA)	Precisely defined size and sequence, sense/antisense strands but relatively insensitive
PCR (ds or ssDNA)	Specific size and sequence, sense/antisense strands, cheap and easy to synthesize

be directly applied to filters as dot blots. This is performed manually where volumes are small, or with the use of vacuum manifolds when larger volumes need to be immobilized (manifolds producing either 'dots' or 'slots' are available). Since it may be difficult to distinguish background from specific signal, the specificity of the probe and the stringency of hybridization and washes need to be rigorously controlled.

Table 2.3. Labelling techniques

Nick translation		Widely used and understood
Random priming		Efficient, but size variable
PCR		Fast, unlimited supply
− 5′ end label	Polynucleotide kinase (isotopic)	Precise, but less sensitive
− 3′ end label	Amino link (non-isotopic)	
	Tailing by terminal transferase	Difficult to control
Photobiotinylation		Quick, cheap, can label any probe type, but no control of probe size and needs special lamp

Restriction enzyme digestion

Non-random fragmentation of genomic DNA is performed by digestion with one (or more) of several hundred restriction endonucleases. These enzymes cut double-stranded DNA at specific DNA sequences; the frequency with which these recognition sites occur in the genome determines the average size of the resultant DNA fragments. Optimization of reaction conditions is sometimes necessary to prevent incomplete digestion or non-specific over-digestion (termed 'star activity').

Electrophoresis

Digested DNA is then electrophoresed in an agarose gel which separates the fragments by size (length). Digests yielding fragments of known size are run simultaneously as a marker so that the positions of the bands detected can be correlated with size. These can be visualized by ultraviolet (UV) transillumination after staining the gel with ethidium bromide.

Transfer from gel to filter

After the DNA has been separated adequately, the gel is treated by methods to facilitate DNA transfer to the filter. If the fragments of interest are large, acid depurination and alkaline hydrolysis are performed to fragment and denature the double-stranded DNA. After neutralization of the alkali, the DNA is transferred to a nylon or nitrocellulose membrane in the presence of a high-concentration salt buffer. Nitrocellulose membranes offer less non-specific background signal, but nylon membranes are mechanically stronger and allow subsequent reprobing of the same target DNA. Positively charged nylon membranes facilitate retention of DNA, but also increase non-specific probe binding. The transfer of DNA between gel and filter can be effected by capillary action known as a 'Southern transfer' (taking 8–16 h), by vacuum suction or by electrotransfer. After transfer, the DNA is fixed onto the filter physically by baking at 80 °C for 2 h, or covalently by UV cross-linking (several minutes).

Pulsed field gel electrophoresis

Methods have been developed recently for separating large DNA fragments (50 kb to 10 Mb). Carefully extracted DNA of high molecular weight is digested in agarose plugs with rare-cutting restriction endonucleases; these detect CG-rich motifs that are under-represented in the genome. The DNA is fractionated in agarose by pulsed field gel electrophoresis (PFGE), using two or more electric fields rapidly alternating in different directions. Southern transfer and hybridization are performd as for DNA of smaller size.

Northern blot

The analogous method for electrophoresis and transfer of RNA is termed northern blotting. Total cellular RNA or poly(A)-containing mRNA is extracted from fresh or frozen samples; the yield from archival human specimens is generally too poor for hybridization analysis. The single-stranded RNA is then electrophoresed in agarose gels containing glyoxal, dimethyl sulphoxide (DMSO), formaldehyde or alkali to denature inter-strand and intra-strand base pairing. Since the RNA is susceptible to degradation by ubiquitous and heat-resistant RNases, these steps are performed in an RNase-free environment.

IN SITU HYBRIDIZATION

ISH basically applies the principle of nucleic acid hybridization to intact tissues or cells (Fig. 2.2). However, the nucleic acid to be detected is retained within the cells and the probe applied to tissue sections or cells mounted on glass slides for hybridization. The cells have to be permeabilized carefully to allow optimum detection. For radiolabels, photographic emulsion is used for detection of bound probe and for non-isotopic labels detection is by immunohistochemistry. The only other significant difference between ISH and filter hybridization pertains to the stringency of the hybridization and washes. Stringency conditions need to be nominally higher for ISH than for comparable probe/target detection on filters (Herrington et al., 1990). This probably reflects the great physicochemical complexity of intact cells in comparison to filters. Permitting, as it does, the detection of nucleic acid targets within structurally intact cellular preparations, ISH is increasingly coming to play an important role in bridging the gap between molecular biology and histopathology.

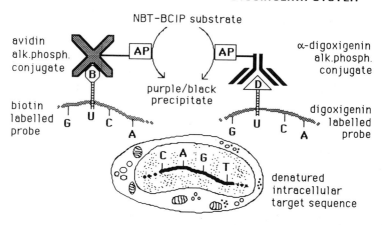

BIOTIN SYSTEM   **DIGOXIGENIN SYSTEM**

Fig. 2.2. Schematic diagram of *in situ* hybridization. See text for details

Advantages of ISH

ISH allows rapid localization of target sequences at tissue, cellular and subcellular levels, and allows correlation of hybridization results with structural features. This, combined with the ability to undertake retrospective investigations of surgical or autopsy formalin-fixed, paraffin-embedded tissues, is of particular advantage in pathology, where the morphological changes can be correlated with presence or absence of a particular nucleic acid target. The use of non-isotopic probe labels and colloidal gold detection has allowed the extension of the technique to the ultrastructural level, permitting detailed analysis of the intracellular localization of target nucleic acids. The facility with which *in situ* hybridization techniques can be combined with immunohistological procedures confers certain advantages: for example, co-detection of nucleic acid sequences and their protein products; unequivocal identification of cell types containing target sequences (by co-detection of cell-type specific antigens); and determination of the proliferative state of cells containing nucleic acid target (by co-labelling with proliferation markers such as the antibody Ki-67). Simultaneous co-detection of more than one target nucleic acid sequence is also possible using different probe labels and detection systems, allowing investigation of questions such as the co-regulation of different mRNA species in the same cell. ISH analysis requires only a relatively limited amount of tissue compared to methods based on nucleic acid extraction, and thus is useful when only small biopsy samples are available. It is also particularly suitable for tissues containing a heterogeneous population of cell types, such as tumours, which not only contain malignant cells but also stromal and inflammatory cells. The fact that a single positive cell may be detected among tens of thousands of negative cells gives the technique a high degree of sensitivity when the target is expressed in a minority population of cells or in a focal distribution. In such instances positive signal would be unlikely to be detected by Southern or northern blotting, due to the 'dilution' effect of the large number of negative cells. Thus, while detection methods based on nucleic acid extraction cannot distinguish between samples in which a minority of cells contain high numbers of targets sequences, and samples in which a majority of cells contain low copy number of the target, ISH permits such differentiation. In experimental situations, the technique permits detection of changes in the pattern and level of gene transcription in response to a variety of external stimuli. The recent development of *in situ* mRNA transcription techniques using labelled nucleotides is an alternative method for localizing mRNA in tissue sections and has significant potential for increasing the sensitivity of *in situ* mRNA detection (Yap *et al.*, 1992). '*In situ* polymerase chain reaction (PCR)' (Murray, 1993) also holds promise for detection of low copy target in tissue sections, although its application in routinely processed tissues requires further validation.

To date, most groups studying gene expression by ISH have used radioactively labelled probes (most commonly ^{35}S-labelled probes), the bound probe being detected by silver grain deposition in a photographic emulsion. Computer-aided image analysis of grain numbers is being used increasingly to give more reliable semi-quantitative results. Radiolabelled probes have the disadvantages of safety hazards, long processing times and relatively poor resolution of signal. Reports on the use of non-isotopic probes for mRNA detection are relatively scarce, partly because of a perceived lack of sensitivity and partly because of difficulties of quantitation of the signal. However, mRNAs have been localized successfully with biotin and digoxigenin-labelled probes and visualized with enzymatic or fluorescent detection systems. Methodological advances will undoubtedly continue to improve the sensitivity of non-isotopic systems, and computerized image analysis systems have been developed which will permit semi-quantification and morphometric analysis of signal from non-isotopic labels. The use of confocal microscopy on non-isotopic probes for *in situ* mRNA analysis is also likely to increase the power of this type of analysis.

Controls

While controls are usually thought to be unnecessary for filter hybridization— because the restriction pattern or size of the target provides specificity in addition to that of the probe and stringency conditions—controls are vital for ISH. Numerous highly convincing artefacts can be generated by ISH. Indeed, on numerous occasions only occasional positive cells or clumps of cells will be seen, leading one to believe the signal must be specific since all the surrounding

Table 2.4 Controls for ISH

Essential

1. No signal with non-homologous probe of same size labelled to same extent
2. Prior digestion with appropriate nuclease (DNase, RNase) removes or significantly diminishes signal
3. Appropriate signal with tissues known to be positive or negative for target
4. Appropriate signal with sense and antisense probes (if available)
5. Appropriate signal with denatured and non-denatured cellular target
6. Probes for unrelated DNA and mRNA to test adequacy of tissue preparation, hybridization, washing and detection conditions, e.g. Y chromosome probe on male tissue, poly-d(T) for whole poly(A) mRNA

Optional

7. Correlation with protein location. However, discordance, particularly with viral nucleic acid and protein, may occur
8. Blockage of signal with unlabelled probe. However, difficult to optimize incubation conditions

---

cells are negative. This may be spurious. Accordingly, a series of controls are essential (Table 2.4).

Of these, the use of unrelated probes of comparable size, labelled to the same extent and used at the same concentration under the same conditions of hybridization, is vital. Similarly, appropriate nuclease pre-digestion should reduce signal intensity, and tissues or cells known to be negative and positive must be used as negative and positive controls respectively. The use of non-denatured tissue or cells is highly advisable for detection of DNA, where no signal should be seen. Conversely, mRNA should be visible even in non-denatured tissues, although signal is usually increased by mild denaturation, probably because of increased accessibility and reduction of secondary structure.

Successful detection of mRNA by ISH is dependent on optimal fixation of target tissues to preserve morphology and mRNA content, and on processing in such a way as to maintain tissue morphology while minimizing RNase degradation of the labile mRNA and permitting adequate access of the probe. Positive controls should always be included to confirm RNA retention in the tissues; possibilities include probing for whole poly(A) mRNA with a poly-d(T) probe, probing for ribosomal RNA, or probing for a 'structural protein' or other mRNA known to be expressed at reasonably high level in the tissue under investigation. In tissues containing plasma cells, the detection of immunoglobulin light chain mRNA (either kappa or lambda) is a useful positive control (see Fig. 2.3). The use of such controls is particularly important in archival specimens where the experimenter has had no control over the processing of the specimens.

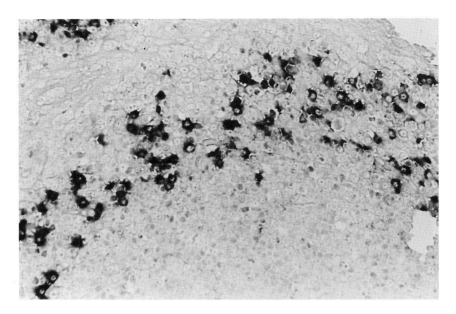

Fig. 2.3. *In situ* hybridization on formalin-fixed paraffin-embedded tissue. Immuno-globulin kappa light chain mRNA in tonsil, detected by digoxigenin-labelled probe

GENETIC ANALYSIS OF MALIGNANCY IN HISTOPATHOLOGICAL MATERIAL

Proto-oncogenes, the normal cellular counterparts of the transforming genes (oncogenes) of mammalian tumorigenic retroviruses, have protein products which are thought to play key roles in the regulation of cellular proliferation and differentiation. Over 100 such human proto-oncogenes have been identified; they include genes coding for polypeptide hormones acting on surface receptors (e.g. the *c-sis* protein product corresponds to platelet-derived growth factor 2 (PDGF-2), receptors for these hormones (*fms* codes for the receptor for colony-stimulating factor 1 (CSF-1)), transducing proteins which convey signals from receptors to the internal milieu of the cell (such as the products of the *ras* family of oncogenes possessing GTPase activity) and nuclear factors affecting transcription from DNA (such as the products of *c-fos* and *c-myc*). Studies both *in vivo* and *in vitro* indicate that mutation or over-expression of proto-oncogenes plays a significant role in the induction and progression of malignancy, and alterations in the expression of one or more proto-oncogenes have been demonstrated consistently in a variety of human tumours (see Table 2.5; also Bishop, 1991; McKenzie, 1991).

Of equal, if not greater, importance in the development of a neoplastic phenotype is the loss of function of anti-oncogenes (or 'tumour-suppressor

Table 2.5. Some oncogenes and tumour-suppressor genes implicated in human cancer (adapted from Bishop, 1991)

Proto-oncogenes	Tumour	Mutation
abl	Chronic myelogenous leukaemia	Translocation
erbB1	Astrocytoma	Amplification
erbB2 (neu)	Breast, ovary, stomach	Amplification
gip	Ovary, adrenal gland	Point mutation
myc	Burkitt's lymphoma	Translocation
	Lung, breast, cervix	Amplification
L-myc	Lung (small cell)	Amplification
N-myc	Neuroblastoma, lung	Amplification
H-ras	Colon, lung, pancreas, breast	Point mutation
K-ras	Acute leukaemias, colon, thyroid, melanoma	Point mutation
N-ras	Genitourinary tract, thyroid, melanoma	Point mutation
ret	Thyroid	Rearrangement

Tumour-suppressor genes	Site	Chromosome
Rb	Reinoblastoma, bone, breast, lung	13q14
p53	Astrocytoma, breast, colon, lung, bone	17p13
WT1	Wilms' tumour	11p13
DCC	Colon	18q21
NF1	Neurofibromatosis type 1	17q11
FAP	Colon	5q21
MEN1	Parathyroid, pancreas, pituitary, adrenal	11q13

genes') which are involved in prevention of uncontrolled cellular proliferation and in promotion of terminal differentiation (Table 2.5). Thus, when the function of these tumour suppressor genes is disrupted, the result is tumour growth. The first recognized tumour-suppressor gene, the retinoblastoma (*Rb*) gene, was postulated by Knudson after analysis of hereditary and sporadic retinoblastomas. The *p53* gene is a tumour-suppressor gene involved in control of the cell cycle (Chapter 6), apoptosis (Chapter 8) and maintenance of genomic stability. Currently abnormalities of this gene are claimed to be the commonest genetic abnormality found in malignant tumours.

Most malignancies probably arise from a combination of abnormalities in both proto-oncogenes and tumour-suppressor genes via a multi-step process; for example, one mutation may result in immortalization of a cell, another in uncontrolled proliferation, a third in invasive ability and a fourth in metastatic potential. Reconstruction of the temporal sequence in which individual genetic lesions accumulate during the progression toward malignancy has been attempted for a number of cancers by analysing tissues from the same anatomical site exhibiting different stages of dysplasia and malignancy. For example, owing to its accessibility and known polyp-to-cancer progression, human colon carcinoma has provided a model of tumour development from normal tissue (Fearon and Vogelstein, 1990) (Figure 2.4). Investigations have shown that individuals with familial adenomatous polyposis (all of whom will develop colorectal cancer) carry inherited mutations in the APC (adenomatous polyposis coli) gene. In addition, up to half of spontaneous tubular polyps carry activated *Ki-ras* oncogenes, and sporadic villous adenomas frequently show allelic deletions of the DCC (deleted in colon carcinoma) gene. Later in the polyp–cancer sequence, invasive carcinomas show *p53* inactivation. Hence, a relationship has been demonstrated between defined stages of tumour progression and specific genetic changes. However, the total accumulation of particular abnormalities rather than their order of acquisition appears to be the crucial feature. Certain proto-oncogenes and tumour-suppressor genes are commonly found to be damaged in a wide variety of different tumours (notably mutations in *ras* oncogenes and loss of the wild-type tumour-suppressor gene, *p53*), while deregulation of other proto-oncogenes appears to be restricted to a particular type of tumour.

Deregulation of normal proto-oncogene or tumour-suppressor activity may occur through a variety of different mechanisms. At the DNA level, sustained or augmented oncogenic activity may occur as a result of (i) point mutations, leading to the production of an abnormal protein product (as is often the case with the *ras* family of oncogenes); (ii) rearrangement or translocation of the proto-oncogene to an abnormal genomic site, disrupting the normal regulatory control, and allowing over-expression and/or production of an altered protein with aberrant function (as happens with *c-abl* when translocated from chromosome 9 to the breakpoint cluster region (*bcr*) on chromosome 22); (iii) amplification of the number of copies of the gene at the DNA level (as frequently occurs with *neu* (*c-erbB-2*) in breast cancer); (iv) deletion of genes, or parts, resulting in inactivation of tumour-suppressor genes through loss of one copy of the gene, followed by mutation in the remaining allele (resulting in reduced expression or inactivation of protein products). In the case of *p53*, mutations at the genomic level can lead to the production of abnormal proteins with transforming effects. A particular mutation at codon 249 of the *p53* gene in a subgroup of hepatocellular carcinomas appears to be related specifically to aflatoxin-induced damage—the first association of an environmental carcinogen with a specific genetic defect. Techniques for identifying such genomic alterations in oncogenes and tumour-suppressor genes by filter hybridization are described below.

Malignancy may also occur without evidence of a genomic abnormality and may result from excessive transcription of mRNA from a specific proto-oncogene, stabilization of mRNA leading to a prolonged half-life, an increase in the rate of protein translation, or a decrease in protein turnover—questions which ISH and immunolabelling techniques are ideally suited to explore. Extraction methods of DNA/RNA analysis and *in situ* analysis of mRNA transcription and protein expression should be viewed as complementary

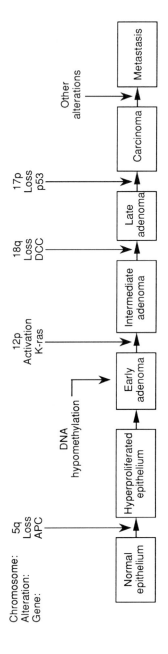

Fig. 2.4. A genetic model for colorectal tumorigenesis. Tumorigenesis proceeds through a series of genetic alterations involving oncogenes (ras) and tumour suppressor genes (on chromsomes 5q, 17p, 18q) (Fearon and Vogelstein, 1990). The order of these changes is not invariant, and accumulation of these changes, rather than their precise sequence, appears more important

rather than competing techniques and when applied in combination can pinpoint the stage at which oncogene abnormality occurs.

GENETIC ANALYSIS OF MALIGNANCY USING FILTER HYBRIDIZATION

As mentioned above, genomic alterations in proto-oncogenes and tumour-suppressor genes are best identified and analysed by filter hybridization techniques. These genomic alterations fall into four main groups as outlined above: gene rearrangements or translocations, amplifications, deletions and point mutations.

Gene rearrangements/translocations

A variety of human malignancies are strongly associated with gene rearrangements. Such malignancies are thought to be the consequence of reciprocal translocation between an activator gene and a proto-oncogene, as mentioned above. Because of the high frequency of the association between tumour type and the specific rearrangement, these changes have great diagnostic utility. To detect and elucidate these rearrangements, new molecular diagnostic techniques based on Southern hybridization (Fig. 2.1) have been developed. These techniques are based on the acquisition or loss of restriction enzyme sites or variation in the size of restriction enzyme fragments resulting from the gene rearrangements. For example, in Fig. 2.5 the immunoglobulin heavy chain rearrangements in Burkitt's lymphoma can be detected by the presence of restriction fragments of aberrant size on Southern blot when compared to DNA from acute lymphoblastic leukaemia and a normal placental control.

The best-understood human malignancies with gene rearrangements are those of the haemopoietic system (Croce, 1991). For instance, the karyotypic marker in chronic myelogenous leukaemia (CML), the Philadelphia chromosome, is a balanced translocation between the long arms of chromosomes 9 and 22 (t9,22; q34,q11)), resulting in the creation of a chimaeric mRNA. The resultant hybrid protein is formed from portions of the *abl* proto-oncogene (the cellular homologue of the transforming gene of the Abelson murine leukaemia virus) and the *bcr* (breakpoint cluster region) gene, whose regulation controls the expression of the bcr–abl protein. Similar, but not identical, chimaeras are formed by translocations occurring in the Philadelphia-positive subsets of acute lymphocytic leukaemia (ALL) and acute myeloid leukaemias (AML). The cytogenetic abnormalities in B and T cell lymphomas have also been studied at the molecular level, revealing rearrangements in the immunoglobulin super-family of genes, including immunoglobulin heavy/light chains and T cell receptor chains. For instance, Burkitt's lymphoma is associated with transloca-tions between the *c-myc* proto-oncogene (8q24) and one of the immuno-

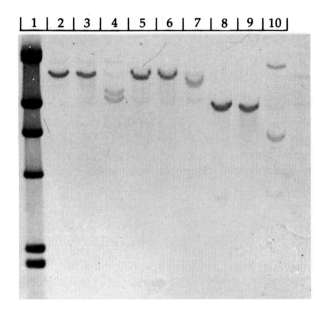

Fig. 2.5. Heavy chain gene rearrangements in Burkitt's lymphoma. Genomic DNA digested by three different enzymes was analysed by Southern blot hybridization with a biotin-labelled probe for the heavy chain J region. Detection of the label was performed using a chromogenic enzymatic reaction. Abnormal restriction length fragment polymorphisms (RFLP) were detected in the Burkitt's lymphoma sample (lanes 4, 7, 10) but not in a case of T cell acute lymphoblastic leukaemia (lanes 3, 6, 9). Placenta DNA was used as a normal control (lanes 2, 5, 8) and a DNA ladder of known sizes was run in lane 1. Photograph courtesy of Oncor, Inc., USA

globulin chains; either the heavy chain (14q32), or the lambda (22q11) or kappa (2p11) light chains, while the translocation found in 90% of follicular lymphomas, t(14;18)(q32;q21), results in enhanced and deregulated expression of *bcl-2*, a gene involved in avoidance of programmed cell death (apoptosis) (Chapter 8). The above examples thus illustrate how molecular analysis has given insight into the pathogenesis of some malignancies. In addition, as mentioned above, many of these translocations are also of great diagnostic specificity.

Gene amplification

Amplification and over-expression of proto-oncogenes occur in many solid tumours. Many of these are from the growth factor receptor family; their over-expression contributes to an up-regulation of the autocrine stimulatory signals for DNA synthesis and cell division. The *c-erbB2* (or *neu*) proto-oncogene, for instance, encodes a transmembrane glycoprotein which shares homology with epidermal growth factor receptor. Protein over-expression correlates with, but is not exclusively due to, gene amplification. Over-expression of c-erbB2 protein in breast cancer has been correlated with a later stage and poorer prognosis in some studies.

Other proto-oncogenes frequently amplified in human tumours are localized to the cell nucleus and are thought to be involved in the regulation of the cell cycle. The c-myc proto-oncogene is a cellular homologue of the transforming gene of avian myelocytomatosis virus MC29, which encodes for a nuclear protein involved in cell cycle regulation. Its amplification has also been reported to be associated with poor prognosis in breast cancer. In neuroblastomas, a related gene, N-myc, is frequently amplified and is also associated with poorer prognosis. In small cell lung carcinoma, amplifications of all three members of the myc family, c-myc, N-myc and L-myc, are seen, although correlation with prognosis is unclear.

The presence and degree of gene amplification in neoplastic cells can be studied by comparing the ratios of the hybridization signals on dot blots of genomic DNA extracted from tumour and non-tumour tissues from the same individual. Constitutive genes which are presumptively not involved in carcinogenesis, and hence regarded as occurring once in every haploid genome, may be used as internal controls. Examples of invasive primary and metastatic breast carcinomas exhibiting amplification of the c-erbB2 proto-oncogene are shown in Fig. 2.6. The myeloperoxidase (MPO) gene was used as internal control.

Gene deletions

The archetype for tumour suppressor genes is the retinoblastoma-susceptibility (Rb) gene in the inherited form of retinoblastoma, a rare bilateral cancer of the eye afflicting the young. In families with the disease, an inactivating point mutation of the Rb allele can be identified. Offspring who inherit this mutated allele in their germline are susceptible to complete inactivation of the gene by a sporadic somatic deletion of the remaining allele, resulting in absence of functional Rb gene product and deregulation of the cell cycle.

Deletion of one of the two parental copies of a gene in the diploid genome can be detected by Southern blot analysis if the two copies differ genetically, i.e. they are heterozygous for a polymorphic locus. The relative amounts of these two alleles can then be compared, and deviations from a ratio of 1 are indicative of either increased copy number of one allele or a diminished copy number of the other. This phenomenon is termed loss of heterozygosity or reduction to homozygosity (Fig. 2.7).

Various types of polymorphisms are used for such analysis. Restriction fragment length polymorphisms (RFLP) are dependent on a mutation in the recognition site of a restriction enzyme, resulting in the gain or loss of restriction site and hence a difference in length. Since such polymorphisms are usually biallelic, there is a maximum theoretical heterozygous rate of 50%.

Case No.	Probe		PT	MT	N

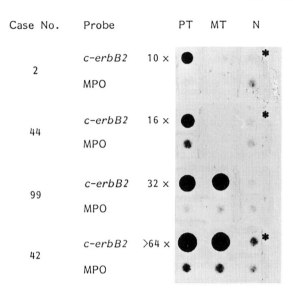

Fig. 2.6. *c-erbB2* amplification in breast carcinoma. Genomic DNA was extracted from archival paraffin blocks of invasive breast carcinomas (primary and secondary tumours) and adjacent non-tumour breast/skin. Dot blots were sequentially hybridized to ^{32}P-labelled probes for *c-erbB2*, and myeloperoxidase (MPO). Quantification of autoradiographic image was performed by densitometry and serial dilution (not shown). PT = primary tumour; MT = metastatic tumour; N = non-tumour tissue

Thus up to only half of individuals and tumour samples are informative and suitable for such analysis. Variable number of tandem repeats (VNTR) are polymorphisms with a varying number of minisatellite insertions, resulting in a larger number of possible alleles. The heterozygosity rate for such polymorphisms is therefore higher and analysis of tumours with VNTR probes yields more informative cases. A third class of polymorphisms called microsatellite (oligonucleotide) repeats is also polyallelic, but the length differences are usually too small to be detected by conventional Southern analysis and are best detected by PCR. Finally, single-base sequence polymorphisms are also best detected by PCR or by single-strand conformation polymorphism (SSCP) (see below).

Karyotyping of interphase chromosomes derived from human cancers has shown interstitial deletions in several chromosomal arms. However, such analysis cannot detect the large majority of small gene deletions, and is also dependent on cultured cells. Using Southern blot analysis of RFLP and VNTR polymorphisms, smaller deletions can be detected. Furthermore, the extent of these deletions may be mapped using adjacent probes along the length of the chromosome. Several tumour suppressor genes have been or are in the process of being identified in this manner, including the *p53* gene on chromosome 17p.

(a)

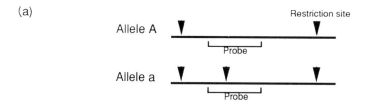

(b)

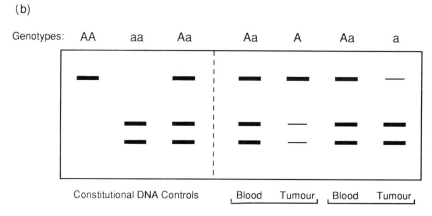

Fig. 2.7. Loss of heterozygosity in cancer detected by Southern blot analysis. Southern blot analysis of a restriction fragment length polymorphism will give two sets of bands in a heterozygous individual, corresponding to the two alleles. Paired analysis of blood (constitutional) and tumour DNA will allow detection of tumours where one of the two alleles is lost. This loss is usually partial due to tumour cell heterogeneity and presence of infiltrating cells (endothelium, inflammatory cells, stroma) within the tumour

The identification of such susceptibility genes would be important in defining subpopulations or pedigrees at high or increased risk of cancer development.

Point mutations

A common mechanism for the activation of oncogenes and inactivation of tumour suppressor genes is a small alteration in the gene sequence which may result in a gene product of increased or decreased length, gross change in the conformation of the polypeptide, small alteration in the functional domains of the protein or abnormal amounts of the normal protein. In a minority of instances, these gene mutations occur in a restriction recognition site and can therefore be detected by Southern blot analysis of RFLP bands. Alternatively, Southern or dot blots may be hybridized at high stringency with allele-specific oligonucleotide (ASO) probes—short DNA probes which are specific for wild-type (normal) or for mutated sequences. A variety of other methods have been developed to screen for this important class of mutations, including single-strand conformation polymorphism (SSCP) and denaturing gradient gel electrophoresis (DGGE). These methods, suitable for analysis of genomic DNA or PCR-amplified DNA, are based on the detection of the altered physicochemical properties of either the single-stranded mutant DNA or the heteroduplexes formed between mutant and wild-type (normal) DNA.

A classical example of oncogenes activated by a point mutation at specific hot-spots is the *ras* family, comprising *Ha-ras*, *Ki-ras* and *N-ras*. Originally found in animal tumours induced by transforming retroviruses, they have been found in many human tumours, including colon, lung, pancreas and thyroid (Table 2.5). The *ras* genes code for a protein, p21, which binds guanine nucleotide phosphates and is involved in cell proliferation. Point mutations occurring in codons 12, 13 or 61 are responsible for activation of the proto-oncogene, essentially by increasing the functional activity of the protein, with resultant increase in cell proliferation.

ANALYSIS OF ONCOGENIC ACTIVATION BY ISH

ISH has been used to investigate human proto-oncogene expression in four main areas: during normal fetal development, in normal tissues, in non-malignant diseases and in malignancy.

Proto-oncogene expression in non-malignant tissues

Studies of human fetal tissues using ISH have indicated that proto-oncogene expression in various organs is developmentally regulated. *C-myc* for instance, is expressed in a tissue- and stage-specific manner in embryonic and fetal tissues and is thought to play an important regulatory role in differentiation pathways. Studies on normal adult tissues show that certain proto-oncogenes continue to be expressed at high levels (for example, *c-fos* and *c-jun* are readily detected in adult skin).

Expression of oncogenes has also been examined in a variety of non-malignant human diseases by ISH; for example, *c-myc* over-expresion has been detected in infiltrating cells in active skin lesions of patients with systemic lupus erythematosus (SLE). Interestingly, the expression of *c-fos* and *c-jun* have been found to be decreased compared with normal adult skin in several dermatoses (e.g. psoriasis) characterized by hyperproliferation and inflammation (Basset-Seguin *et al.*, 1991). This suggests that these proto-oncogenes do not play a crucial role in keratinocyte proliferation *in vivo*, but may play an important role in normal differentiation. ISH for *c-sis* (PDGF-2) mRNA has detected over-expression in the lung epithelial cells as well as macrophages in cases of idiopathic pulmonary fibrosis (Antoniades *et al.*, 1990); this proto-oncogene is a powerful mitogen and chemo-attractant for fibroblasts and smooth muscle cells and is a stimulator of collagen synthesis; thus over-expression may be directly contributing to the abnormal fibroblast prolifer-ation and collagen production seen in this condition.

In situ analysis of proto-oncogene expression in tumours

The activation of over 20 different oncogenes has been studied by ISH in human tumour tissue. In many instances, results have simply confirmed findings of extraction methods. However, the unique ability of ISH to localize oncogene expression in a histological framework is well illustrated by a study on Wilms' tumour published by Pritchard-Jones and Fleming (1991). This tumour is thought to result from loss of function of a tumour-suppressor gene which plays an essential role in normal genitourinary development. The expression of a candidate gene, the WT1 gene (mapping to chromosome 11p13), was studied in tissues from developing fetuses and Wilms' tumour specimens by ISH. It was shown that expression of the WT1 gene was restricted to certain cell types within the developing kidney. Within Wilms' tumours, an abnormal persistence of high levels of expression was seen, but only within neoplastic structures whose counterparts also expressed the gene during normal nephrogenesis, e.g. the blastemal elements. Application of ISH to the study of Wilms' tumour thus represents an opportunity to investigate the relationship between organogenesis and neoplasia.

ISH may also reveal regional differences in expression within different subclones of tumour cells, correlating, for instance, with the degree of dysplasia. Examining the pattern of gene expression within a lesion may thus give a better assessment of its biological importance. Several studies using ISH have revealed over-expression of mRNA for a particular growth factor and its receptor within the same cell populations (e.g. CSF-1 antigen and its receptor (*fms*) in breast cancer; *c-sis* (PDGF-2) and PDGF receptor in astrocytoma), suggesting that autocrine and/or paracrine interactions of growth factors/cytokines and their receptors may play a role in the development and maintenance of some human neoplasias.

Co-localization of oncogene expression in product

Many groups using ISH to detect oncogene mRNA have compared the hybridization results with detection of the oncogene protein product by immunohistochemistry. In many cases, results from the two investigations correlate (e.g. *neu*(*c-erbB2*)) mRNA and protein in breast cancer; Walker *et al.*, 1989), (Fig. 2.8), though in other instances discrepancies exist between the immunolabelling and ISH results (*c-myc* in breast cancer; Walker *et al.*, 1989). In some cases, such inconsistencies may reflect genuine differences between the levels of transcription and translation of oncogenes. However, it should be borne in mind that a variety of technical artefacts can produce the same result.

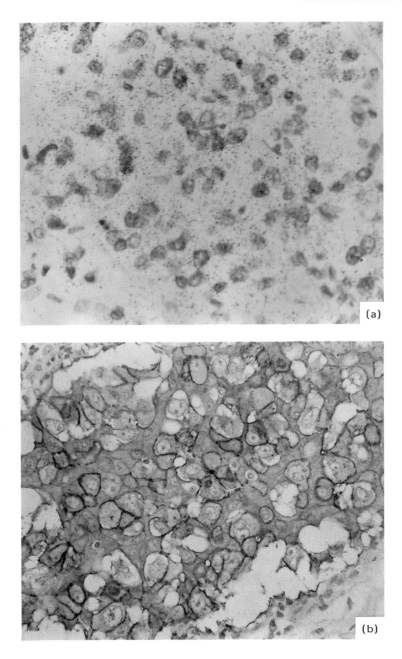

Fig. 2.8. *In situ* hybridization and immunohistochemistry for *c-erbB2*. (a) *c-erbB2* mRNA detected by ^{35}S-labelled riboprobe in a poorly differentiated infiltrating ductal carcinoma of breast. (b) *c-erbB2* protein detection in the same tumour detected by immunohistochemistry. Photographs courtesy of Dr R. Walker, University of Leicester

Many antibodies raised to oncoproteins are non-reactive on formalin-fixed material as a result of masking of the antigenic epitopes by cross-linking, while others demonstrate significant cross-reactivity with other oncoproteins belonging to similar families, or even with cytoskeletal proteins exhibiting partial sequence homology. Conversely, mutations within the coding region of a gene may produce an abnormal protein which is not recognized by antibodies to the wild-type protein. Furthermore, the level of protein detectable by immunolabelling at any given time point will depend not only on the rate of peptide translation, but also on the levels of peptide secretion and/or intracellular peptide degradation. The detection of a certain protein within a cell does not necessarily indicate that it was produced by that cell (it may, for instance, have been made elsewhere and taken up by pinocytosis). Thus the direct detection of mRNA for a specific protein within a cell is far more conclusive evidence that the gene is being transcribed. These reservations aside, the co-detection of oncogene mRNA and protein in the same tissue can be a powerful tool for unravelling the level at which oncogene over-expression is occurring, and the combination is even more powerful when the ISH and immunolabelling techniques are applied consecutively to the same section (see Antoniades et al., 1990). This is no more technically demanding than performing the two procedures on consecutive sections and is particularly readily performed when enzymatic detection of non-isotopic probes is employed.

CONCLUSION

As indicated earlier, a number of studies have shown a statistically significant correlation between activation of particular oncogenes and poor prognosis in a range of different malignancies (e.g. N-myc activation in neuroblastoma, and neu(c-erbB2) activation in cancers of the breast and ovary). Thus, analysis of oncogene status in human tumours is likely to be of increasing clinical significance. It has been suggested that detection of oncogene activation in certain pre-malignant lesions (such as dysplastic Barrett's gastric mucosa in the oesophagus) may be of value in predicting subsequent malignant transformation and that it may help distinguish reactive and neoplastic conditions. Detection of oncogene expression has also been found to be of diagnostic as well as prognostic value; for instance, all cases of Burkitt's lymphoma are associated with c-myc gene translocation and this may help in differentiation from other types of non-Hodgkin's lymphoma.

Analysis of oncogene activation may also assist in the definition of high-risk patients and aid the development of rational therapeutic strategies. While the possibility of repairing genetic damage in cancer is being considered, drugs are being developed whose aim is the inhibition of the effect of particular oncogene products or replacement of missing proteins encoded by tumour suppressor genes. Assessment of oncogene expression may also be used in some instances to monitor the response to specific therapies which aim to down-regulate particular oncogenes (for example, tamoxifen treatment of breast cancer patients appears to affect the expression of several oncogenes (LeRoy *et al.*, 1991) or to target and destroy cells with high oncogene expression. Examination of bone marrow cells for over-expression of *c-myc* oncogenes by ISH has been proposed as an alternative to the use of PCR for the detection of (residual) malignant cells in treated or relapsed patients with acute myeloid leukaemia. Raised levels of certain oncoproteins have been detected in the sera of patients with particular malignancies (see McKenzie, 1991) and assays may be developed which might assist in the identification of patients with occult or residual disease. Antibodies to oncogene products expressed on the surface of cells (such as NEU and EGFR) are also being investigated as a tool for *in vivo* imaging of tumours (for example, by means of a radioactive label attached to the antibody) and could be used to target therapy directly at neoplastic cells.

In addition to the diagnostic and prognostic importance of genetic analysis of malignancy, this approach is also promising to lead to a more complete understanding of the pathophysiological processes underlying tumour growth and spread. Thus, tumorigenesis is thought to be a multi-stage process involving the activation of oncogenes and the inactivation of tumour-suppressor genes, as described above, although the genetic changes necessary to account for the entire phenotype remain largely enigmatic. It should also be mentioned that many mechanisms play a role in inducing or propagating such genetic mutations, including chemical and physical mutagenesis, deficient immunity, increased cell proliferative activity, altered pathways of cell death (apoptosis) and genetic instability.

In summary, the discovery of genetic alterations in human cancer not only furthers the understanding of its aetiology and pathogenesis, but also offers real opportunities for screening populations and pedigrees for genetic predisposition to cancer (e.g., mutations of *p53* or APC), for early detection of cancer (e.g., faecal or urinary analysis of activated oncogenes), for molecular diagnosis (e.g., gene rearrangements in haemopoietic malignancies), for prognostication (e.g., *c-erbB2* amplification in breast cancer) and for suggesting new therapeutic approaches (e.g., drugs targeted at mutant proteins).

REFERENCES

Antoniades HN, Bravo MA, Avila RE *et al.* (1990) Platelet-derived growth factor in idiopathic pulmonary fibrosis. *J. Clin. Invest.*, **86**, 1055–1064.
Basset-Seguin W, Escot C, Moles JP *et al.* (1991) *c-fos* and *c-jun* proto-oncogene expression is decreased in psoriasis: an *in situ* quantitative analysis. *J. Invest. Dermatol.*, **97**, 672–678.
Bishop JM (1991) Molecular themes in oncogenesis. *Cell*, **64**, 235–248.

Croce CM (1991) *Molecular biology of leukemias and lymphomas. Origins of human cancer: a comprehensive review*, pp. 527–542. Cold Spring Harbor: Cold Spring Harbor Lab Press, New York.

Fearon ER, Vogelstein B (1990) A genetic model for colorectal tumourigenesis. *Cell*, **61**, 759–767.

Herrington CS, Graham AK, Flannery DM, Burns J and McGee JO'D (1990) Discrimination of closely homologous HPV types by non-isotopic *in situ* hybridisation: definition of tissue melting temperatures. *J. Histochem.*, **22**, 545–554.

Le Roy X, Escot C, Brouillet JP *et al.* (1991) Decrease of *c-erbB2* and *c-myc* RNA levels in tamoxifen-treated breast cancer. *Oncogene*, **6**, 431–437.

McKenzie SJ (1991) Diagnostic utility of oncogenes and their products in human cancer. *Biochim. Biophys. Acta*, **1072**, 193–214.

Murray GI (1993) *In situ* PCR (Editorial). *J. Pathol.*, **169**, 187–188.

Pritchard-Jones K and Fleming S (1991) Cell types expressing the Wilms' tumour gene (WTI) in Wilms' tumours: implications for tumour histiogenesis. *Oncogene*, **6**, 2211–2220.

Walker RA, Senior PV, Jones JL, Critchley DR and Varley JM (1989) An immunohistochemical and *in situ* hybridization study of *c-myc* and *c-erbB2* expression in primary human breast carcinomas. *J. Pathol.*, **158**, 97–105.

Yap EPH, Martinez-Montero J-C and McGee JO'D (1992) mRNA detection in clinical samples by non-isotopic *in situ* hybridization. In: *Diagnosic Molecular Pathology: A Practical Approach*, eds: Hemington, C.S. & McGee, J.O'D., pp. 187–203. Vol. 1. Oxford: IRL Press.

FURTHER READING

Polak JM and McGee JO'D (1990) *In situ Hybridization: Principles and Practice.* Oxford: Oxford University Press.

Sambrook J, Fritsch EF and Maniatis T (1989) *Molecular Cloning: A Laboratory Manual*, 2nd edn. Cold Spring Harbor: Cold Spring Harbor Lab Press, New York.

3 The Polymerase Chain Reaction: A New Tool for the Histopathologist

L.S. YOUNG

Since its introduction in 1985 the polymerase chain reaction (PCR) has led to a veritable revolution in molecular biology. Given the remarkable simplicity of PCR and the tremendous range of applications for its use, it is no surprise that this technique has had an impact on research similar to that of the discovery of restriction enzymes and of the Southern blot. PCR has been referred to as 'the molecular biologist's photocopying machine' as it allows millions of copies of any specific DNA sequence to be generated within a few hours. The amplified DNA can then be visualized as a distinct band after standard agarose gel electrophoresis and the specificity of detection can be increased by subsequent hybridization or DNA sequencing. The extreme sensitivity of PCR means that a single-copy gene (i.e. β-globin) can be readily detected from extremely small amounts of tissue, and microorganisms present at low levels (i.e. virus infection of one in a million cells) can also be identified. Successful PCR does not require the DNA sample to be pure or of high quality and thus allows analysis of extracts where the majority of the DNA molecules are damaged and/or degraded to such an extent that examination by other molecular biological techniques is impossible. Thus it is possible to study DNA from hair roots, dry blood or semen spots, smears, bone marrow trephines, pre-implantation embryos, paraffin-embedded sections, and even ancient DNA extracted from museum or archaeological specimens. PCR was originally used in the pre-natal diagnosis of gene defects (Saiki *et al.*, 1985) but its use rapidly spread to the analysis of viruses, chromosomal rearrangements and activated oncogenes as well as applications in the areas of forensic medicine and heredity. The PCR has now become a valuable tool in the armoury of the molecular biologist and can be used in a variety of research contexts, including DNA sequencing, analysis of RNA transcription, mutagenesis, development of probes for genetic studies and screening of cDNA libraries. The purpose of this chapter is to review the PCR technique and discuss its application to both frozen and paraffin-embedded histological material.

Molecular Biology in Histopathology. Edited by J. Crocker
© 1994 by John Wiley & Sons Ltd

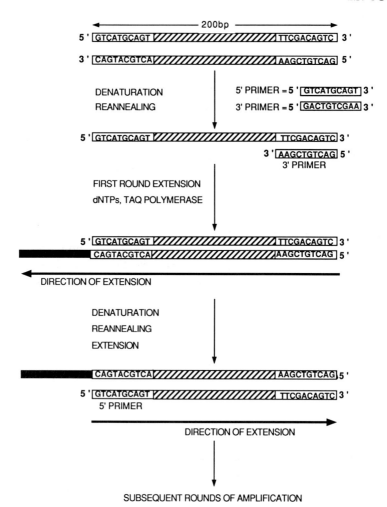

Fig. 3.1. Diagrammatic representation of the polymerase chain reaction. For simplicity the oligonucleotide primers used to amplify the 200 bp fragment are represented as 10mers and the annealing/extension of only one strand is shown. The primers are synthesized to be complementary to DNA sequences on opposite strands of the DNA flanking the fragment to be amplified and thus after denaturation hybridize to their complementary sequences. First-round extension in the presence of deoxynucleotide triphosphates (dNTPs) and Taq polymerase generates newly synthesized DNA strands of indeterminate length, as represented by the solid lines. Subsequent cycles of denaturation, annealing and extension generate fragments whose length is now determined by the boundaries of the primers. Amplification products from previous cycles are used as substrates for subsequent cycles and thus amplified products of a fixed size (i.e. 200 bp) are exponentially produced

PCR: THE BASIC TECHNIQUE

The PCR relies on the ability of DNA polymerases, in the presence of a mixture of deoxynucleotide triphosphates (dATP, dCTP, dGTP, dTTP), to copy a DNA strand using a short complementary DNA fragment as an initiating template. A pair of short DNA fragments referred to as oligonucleotide primers are synthesized to be complementary to DNA sequences on opposite strands of the DNA flanking the fragment to be amplified (see Figure 3.1). The PCR involves repeated cycles of heat denaturation of the DNA, annealing of the primers to their complementary sequences and extension of the annealed primers with DNA polymerase. The orientation of the primers is such that the synthesis of DNA by the polymerase proceeds across the region between the primers. The extension products of one primer can serve as templates for the other primer, so that each successive PCR cycle doubles the amount of DNA synthesized in the previous one. This chain reaction results in the exponential accumulation of the specific DNA target fragment to approximately 2^n, where n is the number of cycles. The size of the amplified DNA fragment is determined by the boundaries of the two primers so that if the chosen region is 200 base pairs (bp) in length, as defined by the 5′ends of the PCR primers, then the amplified product generated during the PCR will be 200 bp (see Fig. 3.1). The amplified product of the predicted size can then be detected by electrophoresis on an agarose or polyacrylamide gel followed by direct visualization with ethidium bromide which lights up the DNA when illuminated with ultraviolet (UV) light (Fig. 3.2). The specificity and sensitivity of the PCR can be further increased by subsequent Southern blotting and hybridization with an oligonucleotide probe internal to the amplifed product.

The PCR requires rapid temperature changes, which can easily be achieved using commercially available thermal cyclers (Fig 3.3). Denaturation of the original DNA sample and the newly synthesized PCR products is used to separate the DNA strands and is usually performed at 90 °C for 30 s. Cooling to temperatures of 37–60 °C for 90 s allows the primers to reanneal to the DNA template. The extension step is then performed by raising the temperature to 72 °C, which stimulates a heat-resistant DNA polymerase (isolated from the thermophilic bacterium *Thermus aquaticus*) called 'Taq polymerase' to copy the DNA template (Saiki *et al.*, 1988). This extension step is stopped by denaturation and the cycle starts over again. The use of Taq polymerase has simplified the PCR procedure by obviating the need to add DNA polymerase every time the newly synthesized DNA strands are denatured, and has therefore made the PCR more amenable to automation. Furthermore, the higher temperatures at which Taq polymerase catalyses the extension step increases the stringency of the PCR, facilitating the amplification of large DNA fragments (over 1 kilobase) which are easily visible on ethidium bromide-stained gels (Fig. 3.2).

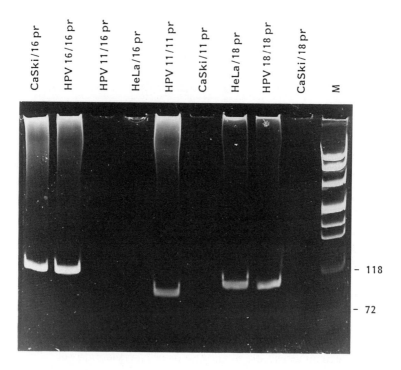

Fig. 3.2. Comparison and specificity of amplification products of human papillomavirus (HPV) types 16, 11 and 18 obtained the polymerase chain reaction (PCR). Polyacrylamide gel electrophoresis and ethidium bromide staining of PCR products obtained with specific oligonucleotide primers and cellular DNA (50 ng) from CaSki (HPV 160-positive) or HeLa (HPV18-positive) cell lines or plasmid DNA (20 ng of either HPV16, HPV11 or HPV18 plasmids). The primers were designed specifically to target the E6 region of the different HPV types and to generate amplified products of different size depending on the HPV type, i.e. 120 bp for HPV16, 90 bp for HPV11 and 100 bp for HPV18. These size differences are clearly seen on the gel. The specificity of the PCR is confirmed by the absence of bands when HPV type-specific primers are used in reactions containing HPV DNA of a different type, i.e. no amplification with HPV16-specific primers in reactions containing HPV11 DNA or HeLa DNA. 11pr, 16pr and 18pr = specific primers for HPV types 11, 16 and 18, respectively; M = DNA size markers

    While the overall specificity of the PCR is determined by the two oligonucleotide primers, other factors influencing PCR specificity include: (i) the times and temperatures of the annealing and extension steps; (ii) the magnesium concentration in the reaction buffer; and (iii) the concentration of Taq polymerase. The optimal conditions for PCR vary with each different pair of primers and are affected by the length and base composition (e.g. GC content) of the oligonucleotides. A simple rule of thumb is that the higher the GC content of a primer the higher the required annealing and extension temperatures. Successful PCR is usually achieved with 20 bp long primers

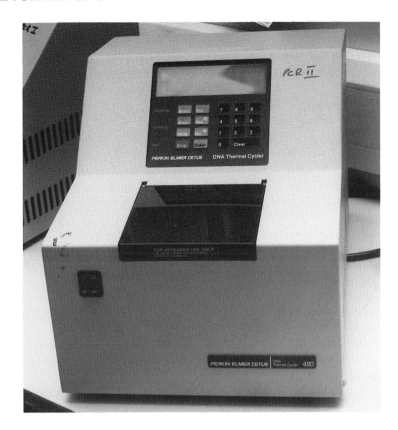

Fig. 3.3. A typical commercial PCR cycling machine

comprising a GC content of 50% in a reaction buffer containing 1.5 mM magnesium.

MODIFICATIONS OF THE BASIC PCR METHOD

More than one set of primers can be used to amplify a specific sequence. The sensitivity and specificity of PCR can be further increased by re-amplifying the PCR product obtained with one set of primers with another set of primers. The second set of primers are complementary to sequences within the amplified product from the first PCR. This technique is referred to as 'nested PCR' and the second set of primers are thus 'nested primers'. However, this procedure is notoriously prone to contamination, a common problem with PCR which will be discussed later. Multiple sets of primers specific either for different genes (e.g. an oncogene such as *c-myc* and a single copy cellular gene such as β-globin) or to multiple regions within a large gene (e.g. the Duchenne muscular dystrophy locus) can be co-amplified in a single PCR. The conditions for this multiplex PCR approach need to be validated carefully to allow equally efficient amplification of the different DNA targets. A very important benefit of PCR is the ability to sequence directly amplified DNA without the need for cloning. A number of kits are now commercially available which use Taq polymerase to sequence PCR products. Amplified DNA is resin-purified and as little as 100 ng of this DNA is directly sequenced in a PCR protocol using one primer—either one of the original PCR primers or a primer internal to the amplified fragment. In this way large amounts of single-stranded DNA are produced and, as the reaction is performed in the presence of appropriate mixes of dideoxynucleotides, simultaneous DNA sequencing is achieved.

The PCR can also be used to generate highly labelled probes for subsequent hybridization to DNA or RNA either on filters or directly on cells and tissues. In this approach the deoxynucleotide triphosphate mix in the PCR buffer is 'spiked' with a labelled nucleotide (usually biotinylated, digoxygenin or radiolabelled dUTP in place of all or a proportion of the dTTP) and the PCR performed as usual (Day et al., 1990). The subsequent amplified product is purified and then used as a probe. This procedure allows specific probes to be generated from complex DNA mixtures such as crudely extracted cellular DNA.

The PCR technique can be easily modified to amplify RNA so that gene expression can be analysed in small amounts of tissue or RNA viruses such as HIV can be detected. In this situation extracted RNA is copied into double-stranded cDNA using the retroviral enzyme reverse transcriptase, and the PCR is then performed on the cDNA copies. Designing PCR primers within different exons of the same gene allows specific amplification across RNA splices, further increasing the specificity of detection (Deacon et al., 1993). The PCR is not quantitative and variability in the amplification efficiencies of different samples even when using the same primers precludes the use of simple strategies for quantitation such as measuring the intensity of the amplified product band. As PCR amplification is an exponential process any small differences in any of the variables that affect the reaction rate (e.g. constituents of the PCR buffer, primer sequences, purity of the DNA sample) will significantly affect the yield of PCR product. In analysing oncogene amplification or virus infection some workers have used two sets of primers in the same reaction: one set to amplify the gene of interest and another set to co-amplify a single copy cellular gene as a reference. However, different sets of PCR primers have inherently different amplification efficiencies which can distort this type of analysis. By far the best approach to quantitate PCR is the strategy involving co-amplification of a competitive template that uses the same PCR primers as those of the target DNA (Gilliland et al., 1990). The competitive template gives rise to a PCR product that can be distinguished from the target DNA by virtue of a restriction enzyme site or a size difference. Target DNA is co-amplified with a dilution series of the competitor DNA of known concentration, and the relative amounts of the two products is determined by direct scanning of the ethidium bromide-stained gel or by incorporation of radiolabelled dNTPs. Because the starting concentration of the competitive template is known, the initial concentration of the target DNA can be determined. This approach is successful but is laborious and requires multiple PCRs for each individual sample which is sometimes precluded by the lack of sufficient material.

SAMPLE PREPARATION

The specificity of the PCR is such that DNA can be amplified from relatively crude extracts. While the technique was pioneered on purified DNA samples, the use of Taq polymerase has obviated the need for DNA prepared using the long-winded phenol/chloroform extraction procedure. Thus cytological speci- mens in the form of buccal/cervical smears or cells isolated from cerebral spinal fluid, urine, pleural or peritoneal effusions can be directly analysed by PCR (Young *et al.*, 1989). Simply boiling the cells is distilled water or PCR buffer is often sufficient but DNA is more efficiently extracted from small numbers of cells by proteinase K digestion followed by boiling. This latter approach can also be used to extract DNA from small tissue specimens or from cryostat sections. Perhaps the major impact of PCR for the histopathologist is the ability to analyse DNA extracted from fixed, paraffin-embedded tissue (Shibata *et al.*, 1988). The possibility of studying preserved material at the molecular level opens up previously unavailable tissue archives for retrospective studies on large numbers of patients. Variable numbers of 5–10 μm thick paraffin- embedded sections, depending on the cross-sectional area of the tissue, can be used to extract amplifiable DNA by dewaxing followed by proteinase K digestion. Paraffin-embedded sections mounted on slides can be treated in the same way. Some workers have experienced problems with PCR on DNA extracted in this way, possibly as a result of the variable fixation procedures employed. Thus, the time taken to process a tissue block and the fixative used (e.g. Bouin's versus formal saline) may influence the overall sensitivity of the PCR.

PCR: A MIXED BLESSING

The ability of PCR to produce millions of copies of a sequence from minute quantities of DNA can easily lead to contamination problems, resulting in false positives. Although false positives can result from accidental sample cross- contamination, the most common and serious source of false positives is the carry-over of PCR products from previous amplifications of the same DNA target. The large number of copies of amplified sequences generated during the

PCR means that extreme caution is required when handling PCR products. To this end, stringent laboratory procedures are needed and the physical separation of PCR preparation from the handling of PCR products is imperative (Kwok and Higuchi, 1989). Thus, the use of a containment area such as a laminar flow cabinet is strongly recommended for setting up the PCR with the use of separate reagents and dedicated pipettes that are kept well away from areas where amplified PCR products are analysed. These precautions may seem excessive but even aerosols, easily generated by uncapping tubes containing PCR products, have been found to be a frequent source of PCR contamination. Meticulous attention to detail is required not only in setting up the PCR but also in all aspects of specimen handling, from sample collection to sample extraction. For instance, where paraffin-embedded material is being processed for PCR, ethanol swabbing of the microtome blade between the cutting of different paraffin blocks is a sensible precaution. Simple extraction procedures that minimize sample handling reduce the possible risks of contamination. In setting up the PCR, strong positive controls should be avoided and the use of negative controls such as reactions containing no template DNA or, where appropriate, DNA negative for the sequence of interest are required in every assay.

False negatives can also be a problem in PCR studies particularly when analysing DNA extracted from paraffin-embedded tissue. In this situation checking that amplifiable DNA has been extracted using primers directed to a single copy cellular gene such as β-globin is a useful test.

APPLICATIONS OF THE PCR

In this section the uses of PCR in the analysis of histological material will be discussed but the wider applications of this technology should not be forgotten. The applications of PCR to cancer research will be emphasized as this highlights the many uses of this technology (Table 3.1).

Table 3.1. Applications of PCR to cancer research

1. Detection of tumour viruses, e.g. HPV in cervical smear cells; HTLV-1 in T cell lymphomas
2. Detection of Ig or TCR gene rearrangements, e.g. clonality of B to T cell lymphomas
3. Detection of chromosomal translocations in haematological malignancies, e.g. t(14;18) in follicular lymphomas; t(9;22) in CML and ALL
4. Detection of point mutations, e.g. activated *ras* genes in pancreatic and colonic carcinomas; α-subunit of Gs in pituitary tumours
5. Detection of tumour-suppressor genes, e.g. Rb in retinoblastoma; p53 in colorectal tumours
6. Detection of amplified oncogenes, e.g. *N-myc* in childhood neuroblastoma; *c-erbB2* in breast carcinoma

GENETIC DISORDERS

Original reports on PCR concentrated on the usefulness of this technique in pre-natal diagnosis of the haemoglobinopathies using amniotic cells or chorionic villi. Since then studies on a variety of single gene defects have been performed, and therefore PCR primers specific for the defective gene have been validated. These studies range from the analysis of the low-density lipoprotein (LDL) receptor gene in familial hypercholesterolaemia to the study of newly identified genes such as the cystic fibrosis transporter protein. Since these inherited mutations are carried in all cells, retrospective analysis can be performed on paraffin-embedded material from any organ. Thus, if a new gene responsible for a single gene defect is identified, studies to evaluate the disease association of that gene can be performed on archival tissue from patients who died many years ago. Similarly, the use of PCR for HLA typing and for identifying disease susceptibility genes can be extended to histological material. The identification of highly polymorphic DNA regions is facilitated by the use of allele-specific oligonucleotide hybridization (Saiki et al., 1986). This approach involves the PCR amplification of a specific region (e.g. the second exon of the HLA-DQα locus) and the subsequent filter hybridization of the amplified product with a labelled allele-specific oligonucleotide probe (e.g. probes specific for HLA-DQα 1.1, 1.2, 1.3, 2, 3 or 4). This technique has been used to study diseases with known or suspected HLA associations such as coeliac disease, type 1 diabetes, rheumatoid arthritis and multiple sclerosis. Such studies require large numbers of patients and thus the ability to analyse archival tissue by PCR will be of great benefit.

RFLP ANALYSIS

Restriction site analysis of PCR-amplified DNA has been widely used to detect genetic variation as determined by restriction length polymorphisms (RFLP). In this approach PCR primers flanking a known restriction site are used to amplify the DNA and the presence of the site is subsequently assessed by treating the amplified product with the appropriate restriction enzyme and determining the size of the digestion products on ethidium bromide-stained gels. In this way, base pair changes, small deletions and insertions can be identified. This approach has been extensively used to detect polymorphisms at various gene loci, including those associated with HLA class II variation and with mutations at specific sites in proto-oncogenes. Modifications of this PCR technique have been applied in the mapping of specific loci. Of particular use in this application are the variable-number tandem repeats (VNTR) which can be detected by PCR and used to generate informative genetic markers. These tandemly reiterated sequences represent a rich source of highly polymorphic markers for genetic linkage, mapping and personal identification. One of the most abundant human repetitive DNA families are interspersed DNA elements

Table 3.2. Detection of viral DNA: sensitivity of available techniques

Technique	Sensitivity
Dot blot	10 gene copies per diploid cell
Southern blot	0.1 gene copies per diploid cell
In situ hybridization	1–10 gene copies per diploid cell
PCR	1 gene copy in 10^5–10^6 diploid cells

of the form $(dC\text{-}dA)_n (dG\text{-}dT)_n$, and these have been widely analysed using PCR with primers either flanking or from within the repeat blocks. The PCR products are then detected after electrophoresis on polyacrylamide gels by staining with ethidium bromide or by autoradiography of products which have incorporated radiolabelled nucleotide during amplification. Thus, PCR offers the benefit of being able to amplify and resolve short repeats, making a large set of previously inaccessible markers available for genetic analysis.

INFECTIOUS AGENTS

An overwhelming number of publications have utilized PCR for the detection of microorganisms in blood, urine, smears and paraffin-embedded material. These studies have examined entities ranging from bacteria and protozoans to a range of different viruses. PCR is particularly useful where other techniques such as serological testing or culture are not applicable (e.g. for detection of mycobacteria or human papillomaviruses). The PCR is at least 10 000 times more sensitive than Southern blotting, being able to detect one gene copy in 10^5–10^6 diploid cells (Table 3.2). In fact, a major problem with this application of PCR is that the extreme sensitivity can result in the detection of clinically insignificant numbers of the organism, making it sometimes difficult to interpret the meaning of a positive result. Furthermore, many workers in this area have been misled by PCR contamination, which has hampered the routine use of PCR in the diagnosis of infectious disease. There is a myriad of papers describing the use of the PCR to detect viruses in clinical specimens. This technique has been widely applied in attempting to ascribe aetiological roles for the herpesviruses, such as cytomegalovirus (CMV), herpes simplex virus (HSV) or Epstein–Barr virus (EBV) and for retroviruses such as human immunodeficiency virus (HIV) or human T cell leukamia virus 1 (HTLV1) in the development of a variety of diseases. The PCR can readily be used to detect genomes of DNA viruses, the integrated proviral genomes of retroviruses and, by inclusion of a reverse transcription step, the genomes of RNA viruses. Transcription of viral genes in cells and tissues has also been analysed using PCR, particularly where only small amounts of material are available. Of particular importance has been the use of PCR to detect viruses associated with certain forms of human cancer. Thus, PCR has been applied to DNA extracted from frozen as well as paraffin-embedded tumours for the detection of viral genomes. An aetiological role for EBV in the development of Hodgkin's disease was originally supported by PCR data, and the role of human papillomaviruses (HPV) in the development of genital neoplasia has been intensively studied using this technique. The studies on HPV serve to emphasize the benefits and pitfalls of PCR analysis and, as such, are worth discussing in more detail.

HPVs are a family of DNA viruses associated with both benign warts and malignant epithelial lesions. Over 60 different HPV genotypes have been identified on the basis of DNA homology, and the different virus types display tropisms either for cutaneous or mucosal epithelium. The viral genomes are small (7.9 kilobases) and structurally related, and can be divided into a 'late' region encoding two structural viral proteins (L1 and L2) and an 'early' region which, in the oncogenic types of HPV, encodes the viral proteins thought to be involved in tumorigenesis. With regard to cervical disease, different virus types are found in pre-malignant and invasive lesions. Thus, HPV types 6 and 11 are predominantly associated with benign lesions (condylomas) and to a lesser extent low-grade cervical intraepithelial neoplasia (CIN), whereas HPV types 16 and 18 are most often found in CIN and invasive cervical carcinomas; HPV types 31, 33 and 35 are less frequently found in cervical tumours. The inability to culture HPVs has prevented the development of *in vitro* assays for detecting infectious virus and has hindered the generation of simple serological tests for HPV infection. Thus, assays for HPV rely on techniques for detection of HPV DNA in infected cells and tissues using the individually cloned HPV types as specific probes in Southern blotting or *in situ* hybridization. While the latter technique is applicable to paraffin-embedded material, it is technically demanding and of varying sensitivity depending on the particular detection system employed (Table 3.2). Thus, the advent of PCR held out great hope that a simple, sensitive and specific technique was now available which could be used for analysis of both extracted DNA and archival material.

The first reports using PCR to detect HPV16 infection in cervical smear samples suggested that the prevalence of infection in both cytologically normal and abnormal specimens was much greater than previously reported, questioning the putative role of HPV16 as the sexually transmitted agent responsible for the development of cervical carcinoma (Young *et al.*, 1989). An alternative explanation for the high prevalence of HPV16 in these studies was the possibility of sample cross-contamination or PCR product carry-over. Subsequently a range of estimates for the prevalence of HPV infection in cervical smears and biopsies have been published (Young *et al.*, 1992). Aside from the concerns over PCR contamination and differences in sample processing, the use of different primers may account for some of this inter-study variation. Researchers have used a number of different primers to detect HPV and these may vary in their sensitivity and specificity, making comparisons between studies difficult. Initial studies used HPV type-specific primers directed to the 'early' region of the viral genome, together with type-specific oligonucleotide probes, i.e. different primers and probes for HPV6, HPV11, HPV16, etc. (see Fig. 3.2). Subsequent studies have made use of general primers directed to highly conserved regions of the HPV genome that are capable of amplifying all genital HPV types, including those of previously unidentified HPV types (Bauer *et al.*, 1991). These so-called consensus primers yield an amplified product of similar size for all the HPVs, and type-specific oligonucleotide probes are then used to identify the specific virus type.

Over the last few years a plethora of papers using PCR to examine DNA infection in a variety of different diseases have been published. The ability to use fixed, paraffin-embedded material in PCR was first demonstrated for the detection of HPV (Shibata *et al.*, 1988). This allows, for the first time, the possibility of examining changes in the prevalence of HPV infection over time and how this relates to the incidence of cervical and other genital cancers. HPV PCR can also be applied to archival Papanicolau smears. The PCR has been used to detect HPV in a variety of lesions, including vulvar and penile warts and carcinomas, anal intraepithelial neoplasia and carcinoma, tumours of the upper respiratory tract as well as tumours of the conjunctiva and oral cavity. Furthermore PCR has been used to exclude a role for HPV in the development of ovarian carcinomas.

There are still a number of problems with the routine use of PCR for the detection of infectious agents. Stringent laboratory procedures are essential to prevent false positives. Standardization of PCR primers is necessary, particularly for epidemiological studies attempting to ascribe an aetiological role for a specific agent in a particular disease. The need for care when using PCR is evidenced by recent work suggesting an association between HPV16 and Kaposi's sarcoma which other groups have been unable to substantiate.

DNA REARRANGEMENTS

PCR can readily be used to amplify across regions of DNA which have undergone rearrangements. In normal B and T lymphocyte development intrachromosomal rearrangements of the immunoglobulin (Ig) and T cell receptor (TCR) genes give rise to the diverse repertoire of the immune system with regard to antigen recognition. These rearrangements are unique to any individual clone and can be used as clonal markers to distinguish the expansion of lymphoid cells associated with neoplastic disease and the polyclonal proliferations characteristic of reactive conditions. Rearranged Ig or TCR genes can be detected in lymphoma biopies or leukaemic cells using standard Southern blotting technology but this is laborious, requiring extraction of DNA from fresh or snap-frozen material and restriction enzyme digestion prior to gel electrophoresis. PCR primers are designed which, while complementary to constant regions, span a region of variable length and sequence generated during DNA rearrangement. Thus, in a clonal population, PCR results in a discrete band of unique size which can be directly visualized on an ethidium bromide-stained gel. In a polyclonal population amplified products from each of the different clones are generated, giving variable sized products which result in a background smear on the gel. An example of this technique is the detection of B cell clones by amplification of the specific sequences encompassing the V–D–J rearrangement of the Ig heavy or light chain loci; a similar approach can be used to detect specific T cell clones by virtue of rearrangements in either the α, β, γ or δ chains of the TCR. In this way PCR has been employed to analyse the clonality of lymphomas in frozen or paraffin-embedded biopsies as well as in lymph node aspirates. The extreme sensitivity of PCR is ideally suited to the detection of disseminated or residual disease. The specificity of this approach can be further increased by tailoring the PCR to detect a specific gene rearrangement. This is achieved by sequencing the PCR product generated from the clonality analysis of the original tumour and designing specific oligonucleotide primers which can then be used to detect that specific malignant clone in future specimens.

The PCR technique described above has been extended to the analysis of chromosomal translocations, some of which occur as accidents during Ig or TCR gene recombination. PCR has been used to analyse the t(8;14) translocation commonly associated with Burkitt's lymphoma, the t(14;18) translocation associated with follicular lymphoma and the 6t(9;22) transloca-tion associated with chronic myeloid leukaemia and acute lymphoblastic leukaemia in frozen or paraffin-embedded tumour tissue, bone marrow, blood and peritoneal washings (Crescenzi et al., 1988). This analysis requires that the chromosomal breakpoints of the particular translocation are known so that one PCR primer from each side of the translocation site can be synthesized. The distance between primers is restricted by the maximum size that the PCR can amplify (around 10 kilobases) and the specificity of the amplified product can be further determined by use of an oligonucleotide probe. In this way only translocated DNA can be detected, giving rise to an amplified product whose size is dependent on the precise location of the breakpoint. This technique is rapid and sensitive and has been applied to the detection of minimal residual disease. However, the prognostic relevance of detecting very small numbers of cells carrying a translocation needs to be assessed. Again, the extreme sensitivity of PCR can lead to the misinterpretation of results, as shown in a number of studies which have used PCR to detect the t(14;18) translocation in Hodgkin's disease. Subsequent work has demonstrated that this translocation is in fact not present in the neoplastic cells of Hodgkin's disease (the so-called Reed–Sternberg cells) but in rare cells present in the activated lymphoid component always found in association with this particular tumour. That PCR can also detect the t(14;18) translocation in reactive lymphoid tissue such as tonsils stresses the care that must be taken in interpreting PCR data with regard to diagnostic or prognostic significance.

ONCOGENES AND ANTI-ONCOGENES

PCR has been used to detect somatic mutations in oncogenes, particularly in the *ras* gene family which are implicated in the oncogenesis of many different human cancers. Single point mutations resulting in the substitution of a single amino acid are sufficient constitutively to activate the *ras* proteins leading to tumorigenesis. These point mutations can occur in all three *ras* genes (N-*ras*, Harvey *ras* or Kirsten *ras*) at codons 12, 13 or 61. Thus, PCR amplification of the individual codons and subsequent hybridization with allele-specific oligonucleotide probes capable of distinguishing between the single base substitutions has been used to examine *ras* activation in a variety of human tumours. An alternative strategy has been the use of primers with a single mismatched base at the 3' end which corresponds to all possible base substitutions found in activated *ras* genes. Individual PCRs are set up which contain each of the 3' mismatched primers with an appropriate upstream primer. As the PCR relies on extension from the 3' end of the primer, amplification will only be possible with the 3' mismatched primer correspond-ing to the point mutation in the sample. Where the mutation in an oncogene or anti-oncogene cannot be predicted, PCR followed by DNA sequencing has been used. This has been widely applied to the analysis of the *p53* tumour-suppressor gene, the most common genetic change detected so far in human cancers. Mutations in *p53* are scattered throughout the *p53* coding sequence but tend to be clustered in exons 5–8. Studies have been published in which each individual exon has been amplified and subsequently sequenced but this is an extremely labour-intensive approach and is not suitable for the analysis of large numbers of specimens. However, samples can be pre-screened for mutations in specific exons using a modification of the PCR. This technique is called single-strand conformation polymorphism (SSCP) analysis and can detect nucleotide substitutions, insertions and deletions in PCR-amplified DNA fragments (Gaidano *et al.*, 1991). SSCP relies on the sequence-specific migration of single-stranded DNA in non-denaturing polyacrylamide gels. PCR products are radioactively labelled during amplification, run out on non-denaturing polyacrylamide gels alongside controls of wild-type DNA amplified with the same primers, and visualization is by autoradiography. In this way the individual DNA strands of the amplified product are observed and size differences between wild-type DNA and sample DNA are an indication of polymorphism. The exons identified to contain mutations by SSCP are then sequenced. This technique has also been used to analyse mutations and deletions in certain exons of the retinoblastoma gene, *Rb1*. Loss of heterozygosity of tumour-suppressor genes or at polymorphic loci can also be detected using PCR by comparing the intensity of the amplified product from wild-type DNA with that of the specimen. A similar approach has been used to detect amplified oncogenes (e.g. N-*myc* in childhood neuroblastoma or *c-erbB2* in breast carcinoma), where the intensity of the PCR product from the specimen is compared with that from a wild-type control or a single copy gene such as β-globin is co-amplified as an internal reference. That all these techniques are applicable to DNA extracted from paraffin-embedded material has again opened up tissue archives so that the significance of oncogene activation or tumour-suppressor gene mutation can be rapidly assessed in large numbers of tumour specimens. As the prognostic implications of these mutations become apparent so the ability to analyse small tumour biopsies using PCR may provide information directly relevant to clinical management.

FUTURE DEVELOPMENTS

New modifications of the PCR which will facilitate the analysis of histological material are continually being published. An example of this is the possibility of using PCR to analyse small numbers of cells that have been histologically identified under the microscope. In this technique specific cell subsets, identified in tissue sections by morphological criteria or by prior immunohistological staining for a particular marker, are protected from UV inactivation by an 'umbrella' in the form of a dot (made with a marker pen) placed physically over the cells of interest (Shibata et al., 1992). Direct UV radiation of the section will cross-link the DNA in all but the protected cells and thus subsequent PCR analysis for virus infection or genetic mutations can be directed to the specific cell subset originally identified. Recently the possibility of combining the sensitivity of the PCR with the cellular localization afforded by in situ hybridization has been investigated. Thus PCR in situ (PISH) or in-cell PCR has been used to detect lentiviral DNA in permeabilized cells, HPV 16 in sections of formalin-fixed cervical and latent HIV infection in lymph node sections from asymptomatic virus carriers (Embretson et al., 1933). This technique is still very much in its infancy and there are a number of theoretical and practical problems with its application. The most encouraging results with PCR in situ have been obtained using multiple primer sets to generate DNA fragments with overlapping cohesive ends in an attempt to increase the retention of PCR products within specific cells in a tissue section. The kinetics of the entry of primers and of Taq polymerase into cells, the efficiency and specificity of in-cell PCR amplification and the prevention of amplified product diffusion have not been thoroughly assessed. Attempts to validate PCR in situ as applied to tissue sections suggest that it is an extremely inefficient process (only 10–100 times amplification versus the 10^6-fold amplification achieved in solution) with significant cell-to-cell variation. Thus, while the theoretical potential of PCR in situ is vast, its practical application in histopathology will have to await the development of more robust and carefully validated technologies.

   The frequency with which papers describing the application of the PCR have appeared in pathology journals over the last few years is testimony to the impact of PCR on histopathology. The ability sensitively and specifically to detect infectious agents and DNA rearrangements of mutated oncogenes/ antioncogenes in fixed, paraffin-embedded tissue or in cytological specimens is of immeasurable value in furthering our understanding of the aetiology of various diseases. In the same way that PCR is becoming widely used for the routine pre-natal diagnosis of genetic diseases, it is envisaged that its application in diagnostic histopathology will also progress to the stage where certain analyses are more readily and reliably performed with PCR than with existing techniques. Just as histopathologists learnt to embrace and apply immunological techniques, so the discipline of molecular biology is now being accepted and modified for use in pathology. In the PCR the histopathology has found a versatile ally with which to examine previously inaccessible material.

REFERENCES

Bauer HM, Greer CE, Chambers JC et al. (1991) Genital human papillomavirus infection in female university students as determined by a PCR-based method. *JAMA*, **265**, 472–7477.

Crescenzi M, Seto M, Herzig GP et al. (1988) Thermostable DNA polymerase amplification of t(14;18) chromosomal breakpoints and detection of minimal residual disease. *Proc Natl Acad. Sci. USA*, **85**, 4869–4873.

Day PJR, Bevan IS, Gurney SJ, Young LS and Walker MR (1990) Synthesis in vitro and application of biotinylated DNA probes for human papillomavirus type 16 by utilizing the polymerase chain reaction. *Biochem. J.*, **267**, 119–123.

Deacon EM, Pallesen G, Niedobitek G et al. (1993) Epstein–Barr virus and Hodgkin's disease: transcriptional analysis of virus latency in the malignant cells. *J. Exp. Med.*, **177**, 339–349.

Embretson J, Zupanic M, Ribas JL et al. (1993) Masive covert infection of helper T lymphocytes and macrophages by HIV during the incubation periods of AIDS. *Nature*, **362**, 359–362.

Gaidano G, Ballerini P, Gong et al. (1991) p53 mutations in human lymphoid malignancies: association with Burkitt's lumphoma and chronic lymphocytic leukaemia. *Proc. Natl Acad. Sci. USA*, **88**, 5413–5417.

Gilliland G, Perrin S, Blanchard K and Bunn HF (1990) Analysis of cytokine mRNA and DNA: detection and quantitation by competitive polymerase chain reaction. *Proc. Natl Acad. Sci. USA*, **87**, 2725–2729.

Kwok S and Higuchi R (1989) Avoiding false positives with PCR. *Nature*, **339**, 237–238.

Saiki R, Scharf S, Faloona F et al. (1985) Enzymatic amplification of β-globin genomic sequences and restriction site analysis for diagnosis of sickle cell anemia. *Science*, **230**, 1350–1354.

Saiki RK, Bugawan TL, Horn GT, Mullis KB and Erlich HA (1986) Analysis of enzymatically amplified β-globin and HLA-DQ α DNA with allele-specific oligonucleotide probes. *Nature*, **324**, 163–166.

Saiki RK, Gelfand DH, Stoffel S et al. (1988) Primer-directed enzymatic amplification of DNA with a thermostable DNA polymerase. *Science*, **239**, 487–491.

Shibata DK, Arnheim N and Martin WJ (1988) Detection of human papilloma virus in paraffin-embedded tissue using the polymerase chain reaction. *J. Exp. Med.*, **167**, 225–230.

Shibata D, Hawes D, Li ZH et al. (1992) Specific genetic analysis of microscopic tissue after selective ultraviolet radiation fractionation and the polymerase chain reaction. *Am. J. Pathol.*, **141**, 539–543.

Young LS, Bevan IJ, Johnson MA *et al.* (1989) The polymerase chain reaction: a new epidemiological tool for the investigation of cervical HPV infection. *Br. Med. J.*, **198**, 14–18.

Young LS, Tierney RJ, Ellis JRM, Winter H and Woodman CBJ (1992) PCR for the detection of genital papillomavirus infection: a mixed blessing. *Ann. Med.*, **24**, 215–219.

KEY BOOKS

Innis MA, Gelfand DH, Sninsky JJ and White TJ (eds) (1990) *PCR Protocols: A Guide to Methods and Applications*. San Diego: Academic Press.

McPherson MJ, Quirke P and Taylor GR (eds) (1992) *PCR: A Practical Approach*. Oxford: IRL Press.

4 Interphase Cytogenetics

J.J. WATERS and S.G. LONG

INTRODUCTION

CHROMOSOME ABNORMALITIES IN CLINICAL MEDICINE

The number of chromosomes in a human metaphase cell first became of clinical interest in the late 1950s, when the findings of a number of workers (Jacobs, Lejeune, Ford and Ferguson-Smith) established a direct relationship between constitutional numerical abnormalities and some clinical syndromes (e.g. between trisomy 21 and Down's syndrome, and between X monosomy and Turner's syndrome). Over the next 15 years techniques were developed that allow the unequivocal characterization of individual chromosomes and their subdivision into visually recognizable zones or bands. A system of nomenclature has been developed to enable accurate description of each chromosome abnormality. This has in turn led to further refinements in the association of clinical syndromes and structural chromosomal abnormalities (e.g. Di George's syndrome and interstitial deletions of chromosome 22).

CHROMOSOME ABNORMALITIES AND MALIGNANCY

Chromosome analysis of leukaemic cells by Nowell and Hungerford in the early 1960s resulted in the first association of an acquired chromosome abnormality, a small marker chromosome which they called the Philadelphia (Ph[1]) chromosome, and chronic myeloid leukaemia (CML). In this case it was established that the chromosomal change was confined to the malignant clone and was not constitutional. Such chromosome aberrations in malignant cells correlate with diagnosis and/or prognosis of the disease, may influence the approach to treatment, and provide a means of monitoring the effectiveness of treatment, including bone marrow transplantation.

Thus, cytogenetic analysis based on identification of chromosomes from metaphase spreads using various banding techniques is now applied in a variety of clinical situations: pre-natal diagnosis, post-natal diagnosis (paediatrics and obstetrics) and malignancy (haematology and oncology).

Molecular Biology in Histopathology. Edited by J. Crocker
© 1994 by John Wiley & Sons Ltd

LIMITATIONS OF METAPHASE CHROMOSOME ANALYSIS

The limitations of this approach are that reasonable numbers of dividing cells are required so that sufficient may be arrested at metaphase, allowing chromosomes to be counted and analysed.

(1) Samples for metaphase cytogenetic analysis must be fresh and in the correct anticoagulant (blood) or transport medium (bone marrow).
(2) For most samples, but not bone marrow, some solid tumours or chorion villous tissue, cell growth must first be established *in vitro*.
(3) Culture times *in vitro* may be lengthy: typically 1–2 weeks for primary amniotic fluid cultures for pre-natal diagnosis.
(4) The cell population that does grow may not be representative of the original sample. In tumour samples this may result in unintentional selection for a highly proliferative subclone or normal cells (e.g. stromal cells). Maternal cells may overgrow fetal cells from tissue in samples taken for pre-natal diagnosis.
(5) Metaphase chromosomes obtained may sometimes be of poor quality. This may result from technical inexperience but some samples (e.g. marrow samples from patients with acute lymphoblastic leukaemia) consistently result in chromosomes of poor morphology which may be tight and fuzzy and difficult to band.

INTERPHASE CYTOGENETICS

Applications initially focused on genome organization and gene mapping, but it soon became clear that isotopic, and more recently non-isotopic *in situ* hybridization (ISH) using chromosome-specific DNA probes, provides a powerful tool for the detection and analysis of numerical and structural chromosome aberrations not only at metaphase but also directly in interphase nuclei. It was for the latter approach that the term 'interphase cytogenetics' was coined by Cremer and co-workers in 1986. It should be noted that 'interphase cytogenetics' does have a longer history if the work on sex chromatin identification in interphase nuclei, by both Barr and co-workers (Barr body) and by Pearson and co-workers (fluorescent Y chromatin), is included. Nevertheless, it is only in the last 7 years with rapid and continuing developments in non-isotopic and in particular fluorescence *in situ* hybridization (FISH), which gives much greater sensitivity and rapidity, combined with the increasing availability of an ever-growing number of chromosome-specific DNA probes, that the potential of interphase cytogenetics in clinical medicine is being realized.

PROBE TYPES

CHROMOSOME LIBRARIES

A chromosome library is a collection of composite probes containing sequences from a given individual chromosome. Collections of specific chromosomes can be obtained using fluorescence-activated sorting. The chromosomes are then digested into fragments of DNA which can then be cloned in a variety of vectors (e.g. plasmids, cosmids, phages). Using this approach the composite probe contains both unique sequences from the specific chromosome as well as sequences that are shared with other chromosomes. When used in FISH experiments the probe hybridizes to the target chromosome and the non-specific sequences are blocked, giving even hybridization along the whole length of the target chromosome. This has been termed 'chromosome painting' (Pinkel *et al.*, 1988) and is most useful in metaphase preparations, but the chromosome library can be seen in interphase nuclei where the chromosomes are spatially arranged in individual domains (Fig. 4.1).

CENTROMERIC PROBES

Alpha satellite DNA is the major class of DNA in the centromeres of human chromosomes. It consists of unique arrangements of 171 bp monomeric repeats present in as many as 5000 copies. Chromosome-specific alpha satellite DNA has now been isolated for most of the human chromosomes and can be used as chromosome-specific probes. Centromeric probes give very bright signals on both metaphase chromosomes and interphase nuclei, allowing the rapid detection of abnormal copy numbers of chromosomes (Fig. 4.1). Unfortunately some centromeric probes bind equally well to more than one centromere and are therefore less useful (e.g. 13/21, 14/22).

COSMID PROBES

Cosmids are vectors specifically designed for cloning large fragments of eukaryotic DNA. The essential components are an antibiotic resistance gene to allow selection of recombinant clones, a plasmid origin of replication site, one or more restriction enzyme sites for cloning and a 'cos' site which allows it to be packaged into phage heads for transfection. They can usually accommodate up to 45 kb of insert DNA. Therefore a cosmid containing a fragment of eukaryotic DNA can be used as a probe in FISH experiments, allowing the detection of single copy sequences in the human genome. The signals are generally quite easily seen on metaphase chromosomes, where they appear as signals on both sister chromatids of individual chromosomes. The signals can also be visualized in interphase nuclei, although the signals are often small and may require sophisticated image analysis equipment to be seen (Fig. 4.1).

Metaphase   Interphase

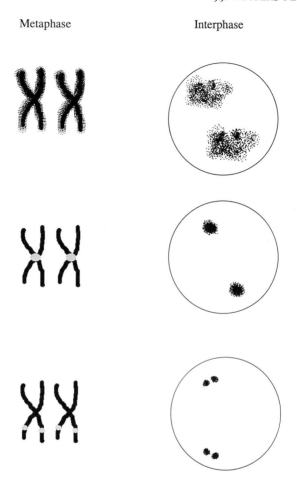

Fig. 4.1. Fluorescence *in situ* hybridization. Top: whole chromosome library; middle: alphoid centromeric probe; bottom: multiple or single-copy, single-site probe

YEAST ARTIFICIAL CHROMOSOMES

Yeast artificial chromosomes (YACs) are vectors in which it is possible to clone even larger fragments than cosmids, up to hundreds of kilobases of DNA. These are therefore more useful in interphase cells as the signals obtained are much brighter, sometimes as bright as centromeric probes. YAC probes are, however, more difficult to generate and use directly in FISH experiments as the yeast DNA interferes with hybridization. This is now being overcome with polymerase chain reaction (PCR)-based approaches to probe generation.

PROBE PREPARATION

The methods used are essentially those described in Chapters 1–3.

SAMPLE PREPARATION

Until recently, FISH studies have concentrated on cytogenetic preparations. Generally, fixed cell suspensions are dropped on to clean glass slides and dehydrated either by baking or dehydrating in ethanol. Other procedures such as exposure to RNase and digestion with proteinases can be used to enhance the signal. More recently, FISH methodology has been applied to other tissues, e.g. blood and marrow smears, solid tissues in paraffin section, etc. Some examples of these are shown in Fig. 4.5, and the reader is referred to the further reading list for detailed method protocols.

IN SITU HYBRIDIZATION

The techniques are essentially those described in Chapters 1 and 2.

INTERPHASE CHROMOSOMES

ORGANIZATION OF CHROMOSOMES AT INTERPHASE

Various studies on the structural organization of whole chromosomes at interphase have shown that:

(1) Each chromosome occupies a defined, localized spatial position ('domain') within the nucleus. This is illustrated when chromosome-specific libraries 'light-up' individual chromosome 'domains' (Fig. 4.1 and 4.5).
(2) Interphase chromosomes are more extended or unwound than their metaphase counterparts.

In addition, one report (Arnoldus *et al.*, 1989) provides evidence for somatic pairing of chromosome 1 centromeres (as defined by probe pUC1.77) in interphase nuclei from paraffin wax-embedded normal brain tissue. In cerebellar samples the expected two spots were observed in 69% of nuclei but in cerebral samples a single large spot was observed in 82% of nuclei. A control probe from chromosome 7 (designated p7t1) gave the expected two spots in cerebellar (83%) and cerebral nuclei (82%).

ORDERING OF GENES USING INTERPHASE CYTOGENETICS

The extended nature of interphase chromosomes has been successfully exploited by Trask (1991), who has demonstrated that interphase nuclei

(a)

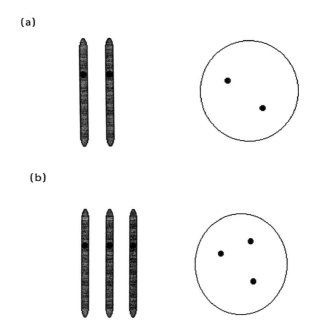

(b)

Fig. 4.2. Schematic illustration of the use of a centromeric probe to detect aneuploidy at metaphase and interphase; (a) two copies of chromosome present, (b) three copies of chromosome present

provide a much more powerful tool than metaphase chromosomes for the ordering of closely linked markers. Thus, while ISH to metaphase chromosomes is now the method of choice for mapping defined DNA sequences, interphase cytogenetics allows the rapid ordering of cosmids which are between 50 kb and 2–3 Mb apart. The minimum inter-marker distance which is resolvable on metaphase chromosomes using ISH is about 1 Mb. This must reflect the greater degree of compaction required at metaphase to maintain a stable chromosome structure.

Some of the simple forms of chromosomal abnormalities which may be screened using interphase cytogenetics are illustrated schematically in Fig. 4.2–4.4. Actual examples are illustrated in Fig. 4.5.

APPLICATIONS

PRE-NATAL DIAGNOSIS: SCREENING FOR ANEUPLOIDY

The most common indication for pre-natal diagnosis is for the detection of chromosome abnormalities and, in particular, trisomy for chromosome 21

(a)

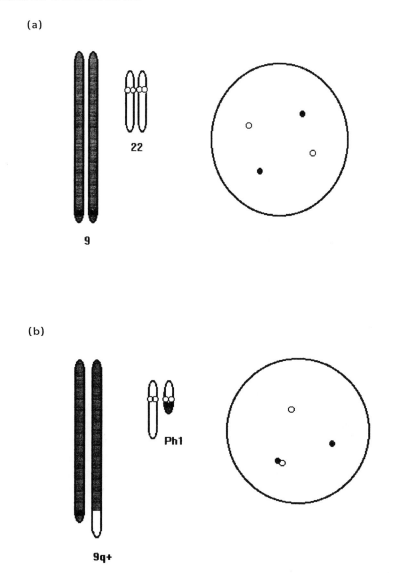

(b)

Fig. 4.3. Schematic illustration of the use of a pair of cosmid probes to detect translocations at metaphase and interphase; (a) no translocation present, (b) translocation present

(Down's syndrome). The test, which involves culturing cells from amniotic fluid samples over a 1–3-week period and analysis of metaphase chromosomes, is labour-intensive and time-consuming.

Interphase cytogenetics performed on uncultured amniocytes or on cells

(a)   (b)

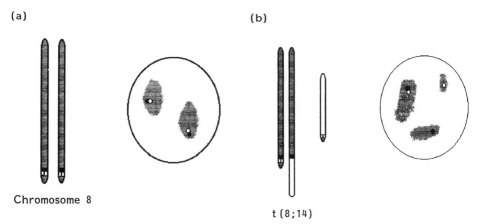

Chromosome 8

t (8;14)

Fig. 4.4. Schematic illustration of the combined use of chromosome-specific libraries and single-copy probes to detect chromosomal rearrangement; (a) closely linked probes on unrearranged chromosome 8, (b) translocation (t8/14) involving breakpoint between closely linked probes on chromosome 8

that have been cultured for relatively short periods (e.g. 1–3 days) offers the prospect of a much faster screening service.

DETECTION OF ANEUPLOIDY USING CENTROMERE-SPECIFIC PROBES

Trisomy for chromosomes 13, 18 and 21 and aneuploidy of the sex chromosomes X and Y make up 95% of all chromosome abnormalities in liveborns which are accompanied by birth defects. ISH with chromosome-specific repetitive centromeric DNA probes reveals their corresponding chromosome regions as distinct spots within interphase nuclei. Thus, with these probes there is a simple correlation between the modal number of interphase spots detected and chromosomal number. Given that in theory the technique has no requirement for cell culture, results for the small number of clinically relevant aneuploidies could be available in 1–2 days rather than several weeks.

This approach has been successfully applied in retrospective studies on uncultured amniocytes and cultured cells (Klinger *et al.*, 1992). Alphoid centromeric probes have proved to be the most popular to date largely because the signals they produce are intense and well localized. However, for the detection of trisomy 21 (clinically by far the most important trisomy), no specific alphoid probe exists. Other probes have been applied, the most promising being a chromosome 21-specific cosmid contig (consisting of two overlapping cosmid probes) which maps to the long arm of chromosome 21 at 21q22 (Zheng *et al.*, 1992).

In practice, the application of these probes to the detection of aneuploidy at

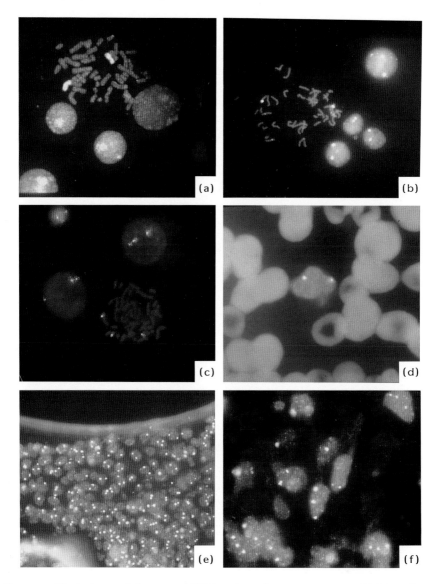

Fig. 4.5. Illustrations of the use of FISH in interphase cytogenetics. (a) Metaphase/interphase cultured lymphocytes. X chromosome library: Biotin–avidin–FITC (BAF) (× 750). (b) Metaphase/interphase cultured lymphocytes. Alphoid 12 centromeric probe (BAF) (× 750). (c) Metaphase/interphase cultured lymphocytes, trisomy 18. Alphoid 18 centromeric probe directly labelled with fluorescein-11-dUTP (× 750). (d) Neutrophil in blood smear. Alphoid centromeric 12 probe directly labelled as (c) (× 750). (e) Tissue section: embryonorhabdomyosarcoma. Alphoid 12 centromeric probe (BAF) (× 200). (f) tissue section: embryonorhabdomyosarcoma. Trisomy 8 with alphoid 8 centromeric probe (BAF) (× 750)

interphase has some drawbacks as a clinical test where reproducibility and robustness are paramount. Most of the cells (>80%) in a mid-trimester amniotic fluid sample are degenerate squamous epithelial cells which are unsuitable targets for ISH. The remaining cells require special pre-treatment to allow adequate exposure of target DNA. Despite these problems, because of the potential saving in time, this area is the focus of much research interest.

SEXING OF HUMAN PRE-IMPLANTATION EMBRYONIC NUCLEI

The molecular basis for some sex-linked recessive conditions is still unknown and selective abortion of male fetuses may be the only option. An alternative approach involves the possibility of sexing preimplantation embryos at the eight-cell stage following *in vitro* fertilization. Removal of one or two cells at this stage does not adversely affect development of the embryo. Sexing of one or two cells using PCR with Y chromosome-specific primers resulted in misdiagnosis of sex in one out of seven fetuses in one study. An alternative approach relies on ISH in fixed interphase nuclei. Dual-colour FISH allows simultaneous detection of both X and Y chromosome-specific sequences. Studies on disaggregated four- to seven-cell human embryos allowed 89% of poor-quality metaphases and 72% of interphase nuclei to be sexed on the basis of the presence of one X and one Y signal or two X signals in the absence of a Y signal (Griffin *et al.*, 1992). Thus, a combination of metaphase and interphase cytogenetics provides a basis for clinical application.

INTERPHASE CYTOGENETICS ON FETAL CELLS IN THE MATERNAL CIRCULATION

A great deal of research effort is currently being spent on methods of analysing the relatively small proportion of fetal cells found circulating in the peripheral blood of a pregnant woman. One approach to detection is to use a nested PCR assay (Lo *et al.*, 1993). Alternatively, enrichment of fetal cells to around 90% has been achieved by fluorescence-activated cell sorting with anti-CD71, anti-CD36 or anti-glycophorin A, or a combination of these three monoclonal antibodies. It then becomes feasible to proceed with an interphase cytogenetics approach to screen for fetal sex or chromosomal aneuploidy.

POST-NATAL DIAGNOSIS

SCREENING FOR ANEUPLOIDY ASSOCIATED WITH CONGENITAL ABNORMALITY

Rapid confirmation of chromosome abnormality may sometimes be required in the neonatal period. Where a chromosomal abnormality such as trisomy 18

Table 4.1. Comparison of the advantages and disadvantages of metaphase karyotyping, flow cytometry and interphase cytogenetics to characterize the chromosomal content of cell populations

	Advantages	Disadvantages
Metaphase Karyotyping	Specific information on chromosome number and structural abnormalities All chromosomes viewed simultaneously	Analysis restricted to a limited number of cells Need for cell culture may select for unrepresentative subpopulations
Flow cytometry	Analysis of a large number of cells Total genomic DNA content profiled	Does not detect small variations in DNA content or structural changes
Interphase cytogenetics	Relatively large number of cells may be analysed (cf. metaphase karyotyping) Specific chromosomal aberrations can be detected No requirement for cell culture Applicable to single cell suspensions, paraffin wax and frozen sections Allows distinction between tetraploid cells at G_0G_1 and diploid cells at G_2M	Decision-making required as to choice of probes to be used (subtractive hybridization techniques may answer this problem)

(Edward's syndrome) is associated with a potentially life-threatening malformation such as tracheo-oesophageal fistula, the outcome of a chromosomal diagnosis may determine whether corrective surgery is performed or not. By using alphoid probes for ISH directly onto fixed blood smears sample preparation time is minimized. By combining this approach with the use of directly labelled probes, removing the need for immunocytochemical detection, results are obtainable within 3–4 h of sample receipt. Such an approach has proven to be diagnostically useful even in the case of post-mortem cardiac blood samples following fetal death *in utero* or after therapeutic abortion, where conventional cytogenetics failed to yield any metaphase spreads (McKeown *et al.*, 1992).

SCREENING FOR CHROMOSOMAL MOSAICISM AT INTERPHASE

The search for chromosomal mosaicism requires a major commitment in time and effort on the part of the cytogeneticist screening metaphase cells. In a diagnostic laboratory, screening 100 cells may represent a practical upper limit.

This represents the exclusion of greater than 2–3% mosaicism within 95% confidence limits. Interphase cytogenetics provides the opportunity to screen, in our experience, at least ten times as many cells in the same amount of time.

SCREENING FOR GENETIC DAMAGE IN INTERPHASE

One recent development has been the use of haploid sperm to screen for genetic damage as a result of environmental exposure to radiation or chemicals. Interphase cytogenetics has been used successfully on sperm to screen for sex chromosome aneuploidy. Using a dual-colour approach, large numbers of cells can be screened simultaneously for X and Y aneuploidy. Screening for structural abnormalities (e.g. deletions) is possible using a combination of centromeric and telomeric probes.

HAEMATOLOGICAL MALIGNANCY

Many haematological malignancies are associated with chromosomal abnormalities. These include chromosomal translocations (e.g. t(9;22)(q34;q11)) in chronic myeloid leukaemia (CML), aneuploidy (e.g. trisomy 8 in myelodysplasia and CML), deletions (e.g. 5q- in myelodysplasia (MDS)) and inversions (e.g. inv (16) in acute myeloid leukaemia (AML)). It is now possible to detect many of these in interphase cells. This has a number of practical uses, such as the detection of residual disease and chimaerism after bone marrow transplantation and the early detection of relapse in leukaemia patients. It may also allow correlation of cytogenetic abnormalities with cell phenotype. Below are some specific examples of interphase cytogenetics in haematological malignancy.

CHROMOSOME TRANSLOCATIONS

Reciprocal translocations in haematological malignancy are relatively common. Although these are most readily detected on metaphase chromosomes they can also be detected in interphase by a number of methods. The t(9;22)(q34;q11) translocation in CML and its consequences are now well understood. Essentially there is fusion of the *bcr* and *abl* genes, usually on the Philadelphia chromosome (Ph'). Bcr/abl fusion can be detected in interphase nuclei using cosmid probes for both the *bcr* and *abl* regions. The probes are labelled with different reporter molecules and visualized in two colours (Fig. 4.3). Overlapping *bcr* and *abl* signals indicate gene fusion (Tkachuk *et al.*, 1990). Chromosome libraries give rather diffuse signals organized into domains in interphase nuclei, and translocation can be inferred when there is disruption of these domains. This approach has been used in combination with single copy probes (phage and plasmids) to detect the t(8;14) translocation in Burkitt's lymphoma (Reid *et al.*, 1992). The technique used a chromosome 8 library and a pool of phage clones spanning a 170 kb region either side of the breakpoint near the MYC oncogene. Tumour cells were recognized when the phage signals from the region distal to the breakpoint fell outside of the interphase chromosome 8 domain (Fig. 4.4).

ANEUPLOIDY

MDS (and acute myeloid leukaemia) are commonly associated with mono-somies 5 and 7 and trisomy of chromosome 8. These numerical abnormalities are easily detected in interphase nuclei by using alpha satellite pericentromeric probes. The single most common chromosomal abnormality in chronic lymphocytic leukaemia is trisomy of chromosome 12 and can readily be detected on interphase cells with an alpha satellite probe (Losada *et al.*, 1991). This technique can also be applied to previously stained cells, e.g. bone marrow smears, and to analyse morphologically suspect cells.

MIXED CHIMAERISM

Bone marrow transplantation (BMT) is the only hope of cure for many malignant haematological conditions. When the donor and recipient are of opposite sex interphase, FISH with probes for X and Y centromeric regions can be used to assess the degree of mixed chimaerism, which may have prognostic implications. The analysis of interphase cells avoids relying on metaphase chromosomes which can be misleading after BMT, when normal donor cells will often divide at a more rapid rate than residual malignant cells (Anastasi *et al.*, 1991). Interphase FISH may prove to be useful in the detection of minimal residual disease and has some advantages over PCR, which has now been extensively used for this purpose after BMT. FISH is a more easily quantifiable technique than PCR and may also allow correlation of residual clonal cells with cell phenotype.

CONCLUSION

ISH in interphase cells has already found widespread application in pathology. ISH gives rapid and statistically more reliable information about chromosomal numerical and structural abnormalities compared with metaphase cytogenetic methods and gives more precise information than flow cytometry.

FISH allows multiple labelling systems to be used to visualize two or more target DNA sequences in the same metaphase or nucleus, or to visualize different RNAs in the same cell or in different subpopulations of cells in the same tissue section. Computer-aided visualization, enhancement, storage and information-processing are other rapidly growing areas which allow the routine detection of single copy sequences. Finally, approaches combining a variety of immunohistochemical techniques and ISH will find increasing application in diagnostic and experimental pathology.

ACKNOWLEDGEMENTS

We wish to thank our collegues Dr Don Cardy for supplying Fig. 4.5C, Dr Nicola Bienz for Fig. 4.5D and Dr Felix Niggli for Fig. 4.5E and F. We would also like to thank Professor Maj Hultén for her support and encouragement. Dr Simon Long was formerly a Sheldon Clinical Research Fellow with the West Midlands Regional Health Authority. This work has also been supported by the Leukaemia Research Fund.

REFERENCES

Anastasi J, Thangavelu M, Vardiman JW et al. (1991) Interphase cytogenetic analysis detects minimal residual disease in a case of acute lymphoblastic leukaemia and resolves the question of origin of relapse after allogeneic bone marrow transplantation. Blood, 77, 1087–1091.

Arnoldus EPJ, Peters ACB, Bots GTAM, Raap AK and van der Ploeg (1989) Somatic pairing of chromosome 1 centromeres in interphase nuclei of human cerebellum. Hum. Genet., 83, 231–234.

Cremer T, Landegent J Bruckner A, Scholl HP, Schardin M, Hager HD, Devilee P, van de Ploeg M (1986) Detection of chromosome aberrations in the human interphase nucleus by visualization of specific target DNAs with radioactive and non-radioactive in situ hybridization techniques: diagnosis of trisomy 18 with probe L1.84. Hum. Genet., 74, 346.

Griffin DK, Wilton LJ, Handyside AH, Winston RML and Delhanty JDA (1992) Dual fluorescent in situ hybridization for simultaneous detection of X and Y chromosome specific probes for the sexing of human preimplantation embryonic nuclei. Hum. Genet., 89, 18–22.

Klinger K, Landes G, Shook D et al. (1992) Rapid detection of chromosome aneuploidies in uncultured amniocytes by using fluorescence in situ hybridization (FISH). Am. J. Hum. Genet., 51, 55–65.

Lo DY-M, Patel, P, Baigent CN, Gillmer DG, Chamberlain P, Travi M, Sampietro M, Wainscoat JS, and Fleming KA (1993) Prenatal sex determination from maternal peripheral blood using the polymerase chain reaction. Hum. Genet., 90, 483–488.

Losada AP, Wessman M, Tiainen M et al. (1991) Trisomy 12 in chronic lymphocytic leukaemia: an interphase cytogenetic study. Blood, 78, 775–779.

McKeown CME, Waters JJ, Stacey M et al. (1992) Rapid interphase FISH diagnosis of trisomy 18 on blood smears. Lancet, 340, 495.

Pinkel D, Landegent J, Collins C et al. (1988) Fluorescence in situ hybridization with human chromosome specific libraries: detection of trisomy 21 and translocations of chromosome 4. Proc. Natl Acad. Sci. USA, 85, 9138–9142.

Ried T, Lengauer C, Cremer T et al. (1992) Specific metaphase and interphase detection of the breakpoint region in 8q24 of Burkitt lymphoma cells by triple colour fluorescence in situ hybridization. Genes Chromosome Cancer, 4, 69–74.

Tkachuk DC, Westbrook CA, Andreeff M *et al.* (1990) Detection of *BCR–ABL* fusion in CML cells by two color *in-situ* hybridization *Science*, **250**, 559–562.

Trask B (1991) Fluorescence in situ hybridization: applications in cytogenetics and gene mapping. *Trends genet.*, **7**, 149–154.

Zheng YL, Ferguson-Smith MA, Warner JP *et al.* (1992) Analysis of chromosome 21 copy number in uncultured amniocytes by fluorescence in situ hybridization using a cosmid contig. *Prenat. Diagn.*, **12**, 931–943.

FURTHER READING

Anastasi J (1991) Interphase cytogenetic analysis in the diagnosis and study of neoplastic disorders. *Am. J. Clin. Pathol.*, **95** (Suppl. 1), S22–S28.

Hopman AHN, van Hooren E, van de Kaa CA, Vooijs PGP and Ramaekers FCS (1991) Detection of numerical chromosome aberrations using in situ hybridization in paraffin sections of routinely processed bladder cancers. *Modern Pathol.*, **4**, 503–513.

Lichter P and Cremer T (1992) Chromosome analysis by non-isotopic in situ hybridization. In: Rooney DE and Czepulowski BH (eds), *Human Cytogenetics: A Practical Approach* Vol. I (Constitutional Abnormalities), 2nd edn, pp. 157–173. Oxford: IRL Press.

Poddighe PJ, Ramaekers FCS and Hopman AHN (1992) Interphase cytogenetics of tumours (Review article, Chromosome pathology). *J. Pathol.*, **166**, 215–224.

Warford A and Lauder I (1991) In situ hybridization in perspective. *J. Clin. Pathol.*, **44**, 177–181.

Westbrook CA (1992) The role of molecular techniques in the clinical management of leukaemia. Lessons from the Philadelphia Chromosome. *Cancer*, **70**, No. 6 (Suppl.), 1695–1700.

5 DNA Flow Cytometry

R.S. CAMPLEJOHN

INTRODUCTION

DEVELOPMENT AND PROPERTIES OF FLOW CYTOMETERS

Flow cytometry has developed over the past 30 years from initial attempts to count and size particles. The first cell sorter was described in 1965 and multiparameter machines measuring two fluorescence wavelengths were described around 1970. An excellent summary of the historical development of flow cytometers is given in the first chapter of the book edited by Melamed *et al.* (1990).

Currently two main categories of flow cytometer are commercially available. Firstly, there are simple analytical bench-top machines which typically can measure five simultaneous parameters on single cells or nuclei as they pass through the cytometer. Three of these parameters are related to the fluorescence characteristics and two to the light-scattering properties of the cells. The second major category of flow cytometer is that of large cell sorters. These machines not only measure five or more parameters on particles passing through them but they can also physically sort particles with a desired set of properties into separate containers for further study. Many cell sorters have the ability to use multiple lasers for excitation of cellular fluorescence; this allows a wider variety of fluorochromes to be used. There are just now appearing a new generation of cell sorters which are smaller and cheaper to run. They are compact 'bench-top' machines with the ability to sort. Such machines may bring cell sorting within the financial range of an increased number of laboratories.

Bench-top analytical flow cytometers are of most interest to the routine pathology laboratory and further discussion will be restricted largely to the use of such machines. Modern devices of this type generally use a small (15 mW) air-cooled argon-ion laser tuned to 488 nm (i.e. producing blue light). Such machines are simple to use for routine applications. Conversely, their use is restricted in research by their often fixed optical set-up and single laser line. This light beam (usually set at a wavelength of 488 nm), excites fluorescent dyes used to stain cellular constituents of interest, such as DNA. The

Molecular Biology in Histopathology. Edited by J. Crocker
© 1994 by John Wiley & Sons Ltd

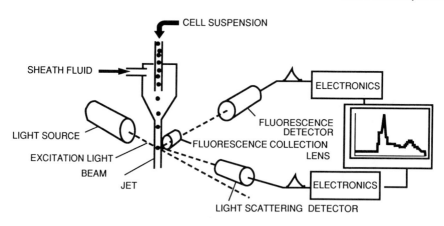

Fig. 5.1. Schematic representation of a flow cytometer. This is a very simplified diagram showing only one fluorescence and one light-scatter detector. Modern flow cytometers would generally have at least three fluorescence and two light-scatter detectors

fluorescence from one or more dyes is collected, split into its different components (if more than one dye is involved) by a series of mirrors and lenses and measured by detectors. Signals from the detectors are then processed electronically into a form suitable for storage and analysis on a computer (see Figure 5.1).

In this chapter, discussion will be restricted to applications of flow cytometry that have relevance to a histopathology department dealing with clinical material. However, it is worth commenting on the breadth of applications of flow cytometry in medicine and in cell and molecular biology. Techniques are now available to measure flow cytometrically a wide variety of cellular constituents and properties. For example, DNA, RNA and protein content can be measured and, with the aid of monoclonal antibodies, the presence of specific proteins and other molecules in individual cells can be determined. Intracellular pH, changes in intracellular calcium levels and various aspects of cellular metabolism can be assessed. As we will mention briefly later, flow cytometry also has clinically relevant uses in studying genetic changes at the chromosomal level in various pathological conditions.

GENERAL CHARACTERISTICS OF FLOW CYTOMETRY

Flow cytometry has the advantages of speed and statistical precision. Typically 10 000–100 000 cells or nuclei can be scanned in a few minutes or less. Given suitable staining conditions, measurements of fluorescence in individual particles are related to the amount of the stained substance in or on the particle, i.e. measurements are quantitative. Multiple parameters can be measured

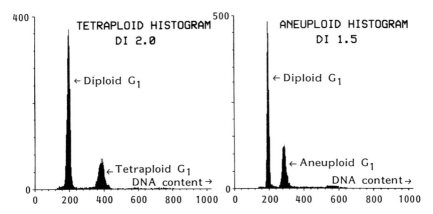

Fig. 5.2. Both of these plots were obtained from needle aspirates of breast carcinomas. The right-hand plot shows a DNA aneuploid histogram. In this case the aneuploid G_1 cells contain 50% more DNA than do the diploid G_1 cells; this tumour is thus said to have a DNA index (DI) of 1.5. The left-hand plot illustrates data from a tumour containing a tetraploid stem line; these cells have a DI of 2.0. The tetraploid cells have S and G_2/M phases (see also Fig. 5.5)

simultaneously on individual cells. This can be a very useful attribute of the technique.

No technique is without its disadvantages, however, and flow cytometry is no exception. Among the disadvantages of flow cytometry is that a relatively expensive machine is required. In addition, when studying solid tissues, it is entirely necessary to disaggregate the tissue into a suspension of single cells or nuclei. Some solid tissues are difficult to disaggregate and in all cases tissue morphology is lost.

PRINCIPLES OF DNA FLOW CYTOMETRY

The earliest flow cytometrically measured DNA histograms, which showed clearly defined G_1, S and G_2/M phases of the cell cycle, seem to have been produced in the late 1960s. Since then, a vast number of studies have been published, in which DNA content of clinical material has been measured. In this chapter the discussion will be restricted to cancer-related topics.

Two main parameters can be calculated from DNA histograms. The first of these relates to the presence of cells with abnormal amounts of DNA, so-called DNA aneuploid cells (see Fig. 5.2). The extent of the deviation in DNA content from normal is defined by the DNA index, as described in the legend to Fig. 5.2. Early studies using both static and flow cytometry tended to show that tumours containing DNA aneuploid cells had a worse prognosis than those containing only DNA diploid cells. The second major parameter to be gleaned from DNA histograms is a crude index of proliferative activity. In general,

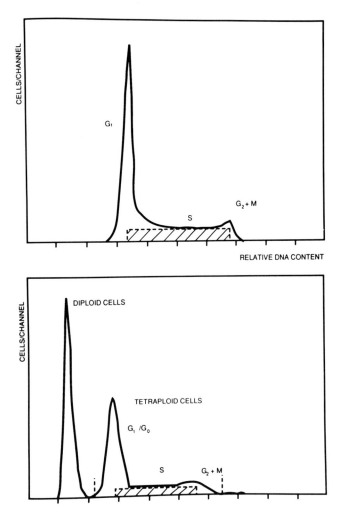

Fig. 5.3. The top panel represents a DNA diploid histogram. The S phase fraction (SPF) is calculated in our laboratory by fitting a rectangular area to represent S phase. The lower panel represents a tetraploid histogram; in this case, SPF for the tetraploid cells is calculated by fitting a rectangle to represent the tetraploid S phase. SPF is quoted as the number of tetraploid S phase cells divided by the total number of tetraploid cells × 100. This modification of the Baisch technique can be applied to most aneuploid histograms (for exceptions see Fig. 5.6)

where proliferative activity is high, there will be many cells in the S and G_2/M phases of the cell cycle and, conversely, when poliferative activity is low there will be few such cells. Thus, the percentage of cells in the S phase (the S phase fraction or SPF) or the combined percentage of cells in the S + G_2/M phases are often used as crude indices of proliferative activity (Fig. 5.3).

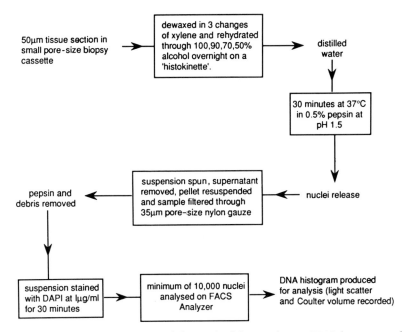

Fig. 5.4. Schematic representation of the method for producing DNA histograms from 50 μm paraffin-embedded tissue sections. In this diagram the use of DAPI as the DNA fluorochrome is described but DAPI can be replaced by PI plus RNAse

METHODOLOGY

THE USE OF PARAFFIN-EMBEDDED MATERIAL

A major stimulus to clinical flow cytometric DNA studies came in 1983, with the publication of the paper by Hedley and co-workers. This study described a method of disaggregating thick paraffin-embedded sections using pepsin, so as to produce nuclear suspensions suitable for the performance of DNA flow cytometry. This method has been modified slightly (Fig. 5.4) and much of it has been automated to increase the convenience and productivity of the method in routine application.

ADVANTAGES AND DISADVANTAGES OF PARAFFIN-EMBEDDED MATERIAL

The use of paraffin-embedded material has some logistic advantages over the use of fresh tissue for clinical flow cytometric studies. For example, large retrospective series of patients, for whom clinical follow-up is already available, can be investigated. This is particularly useful for breast cancer, as an example, where 10 or more years follow-up is ideal before conclusions can be made relating to survival. Also, if one wishes to study rare lesions, the use of archival material enables a reasonable series of patients to be collected. DNA flow cytometric studies from paraffin are also logistically easy for large multi-centre studies in which the flow cytometry is to be performed centrally; 50 μm sections can be sent in any convenient sealed container (small plastic tubes are ideal) by post to the institution in which flow cytometry is to be carried out.

What then are the disadvantages of using paraffin-embedded material? One of the most common problems quoted is that the quality of the histograms produced is poorer related to those obtained with fresh tissue; this finding was noted by Hedley *et al.* (1983) in their original study. Others have not found this poorer quality in a comparison of fresh and paraffin-embedded lymph node biopsies (Camplejohn *et al.*, 1989) but it is fair to state that most workers who have done comparisons have found paraffin-embedded tissue to yield poorer quality data. One factor which influences quality is the nature of the fixation to which archival material has been subjected. It is well known that certain fixatives (such as Bouin's and those containing mercury) lead to poor-quality DNA profiles. In addition, with fixation in formalin-based fixatives, certain factors such as over-fixation, the use of acid formalin and high temperatures militate against good-quality data. Overnight fixation in neutral buffered formalin at 4 °C or the use of formal phenol are ideal protocols. However, if one wishes to perform retrospective studies, then one is forced to use material fixed by whatever protocol was in use during the period of interest. It is often the case that within one institution variations in methods of fixation occur quite frequently and a given study might encompass samples fixed in a variety of ways. Fortunately, most routinely used formalin-based fixation schedules allow the majority of samples to be analysed but success rates do vary between 50% and 95% depending upon the precise protocol employed. Little can be done, after the event, to overcome the effects of inappropriate fixation in archival material.

In general, when DNA flow cytometric results have been compared from fresh and paraffin-embedded material, agreement in terms of ploidy has been good. The situation in relation to SPF is more variable but the majority of authors have obtained reasonably good agreement (see, for example, Camplejohn *et al.*, 1989).

NEEDLE ASPIRATES AS A SOURCE OF MATERIAL FOR DNA FLOW CYTOMETRY

In view of the aims of this book, much of the foregoing describes the application of DNA flow cytometry to paraffin-embedded material. However, needle aspirates can provide another practical source of material for clinical measurement of cellular DNA content. Aspirates may be taken from some tumour types (e.g. breast and lymphoma) *in situ* and can be obtained from virtually all types of tumour after excision (Vindelov and Christensen, 1990). If a skilled operator takes the aspirates, enough cells can be obtained in most cases to provide good-quality DNA profiles. A simple DNA staining protocol can be applied by permeabilizing cells with detergent. Alternatively a staining kit, which can be bought commercially, is available based on the method of Vindelov. In the author's laboratory, a high success rate (around 95%) has been achieved in obtaining analysable DNA profiles from breast carcinomas aspirated *in situ*. This is despite the fact that only a part of an aspirate taken for diagnostic cytology was provided. For prospective DNA measurements within a single institution, needle aspiration may be the method of choice.

STAINING AND FLOW CYTOMETRIC MEASUREMENT

A number of fluorochromes are available which can be used to stain DNA quantitatively. The commonest dye, which is suitable for use with modern bench-top flow cytometers employing a small laser tuned to 488 nm, is propidium iodide (PI). PI intercalates between base pairs of both double-stranded DNA and RNA. Thus, when PI is used to measure DNA content, RNase should be used to remove double-stranded RNA. As with all DNA dyes, PI should be used at a concentration which leads to the dye being present in modest excess. In this way, all available binding sites on the DNA are bound to PI and quantitative staining is achieved. Some bench-top flow cytometers employ a mercury arc lamp as the light source, and in this case, DNA fluorochromes which excite at wavelengths other than 488 nm can be used. A dye with the long name 4,6-diamidino-2-phenylindole-dihydrochloride, which mercifully is shortened in everyday parlance to DAPI, can be used. This dye is DNA-specific and does not require RNase pre-treatment. The choice of DNA stain should not influence the nature of the results obtained in routine clinical studies and, indeed, exceedingly good agreement between data obtained with PI and DAPI has been found (Camplejohn *et al.*, 1989). With nuclei from paraffin sections and cells/nuclei fixed in alcohol, DNA dyes such as PI will penetrate into the nucleus. If intact, unfixed cells are used they must first be permeabilized, as described above for cells obtained from needle aspirates. For detailed protocols as used in the author's laboratory see Camplejohn (1992).

Whatever DNA dye and type of flow cytometer are used, it is usual to measure DNA content in at least 10^5 nuclei per sample. This can be done very rapidly with typical flow rates of 100 or more cells per second. Thus, each profile requires only 1–2 min to acquire. The data obtained are then converted into digital form for storage and analysis on a computer. The data are acquired and stored in what is termed 'list mode', meaning that information on several parameters (e.g. forward and $90°$ light scatter, peak and area of DNA fluorescence) is obtained and stored in a correlated form, so that for every nucleus the different parameters can be related to that individual particle.

DATA ANALYSIS

Perhaps the most complex and vexatious aspect of DNA flow cytometry is data analysis. Some attempts have been and are being made, for example, by the International Society for Analytical Cytometry, to standardize procedures and nomenclature. Nevertheless, many different methods are used to analyse data, often with the aid of a variety of computer algorithms. No attempt will be made here to review the literature on this topic. However, it is clear that differences in methods of data analysis are a significant factor in the different conclusions reached in published work concerning the clinical significance of DNA flow cytometric studies. In the present chapter the methods used in the author's laboratory will be described, with some explanation as to why we use the methods we do.

DNA ploidy

Firstly, the measurement of DNA ploidy will be considered. The basic principles by which DNA aneuploidy is recognized and by which DNA index is calculated are widely accepted (Fig. 5.2). With fresh tissue samples, such as those obtained from needle aspiration, external DNA standards can be used as a means of confirming the DNA ploidy status of a given DNA peak in a test sample. The most accepted way of doing this is to use human peripheral blood cells, although nucleated red cells from fish and birds are also sometimes used. Such external DNA standards cannot be used with paraffin-embedded material, as was pointed out in the original paper by Hedley *et al.* (1983), and has been confirmed many times since. The reason for this is the variability in staining intensity of DNA in nuclei from different paraffin blocks, resulting from differences in fixation and processing. There are a number of publications in which erroneous conclusions concerning DNA ploidy have resulted from the inappropriate use of 'external standards' in studies of paraffin-embedded material. Thus, with DNA histograms from paraffin-embedded blocks, the convention is that where two G_1 peaks are seen the left-hand peak is considered to be DNA diploid and the DNA index (DI) of the other peak is calculated accordingly. This should not lead to the failure to recognize DNA aneuploidy as all samples contain reference DNA diploid cells (lymphocytes, endothelial cells, etc.), but it will lead to a small percentage of samples being wrongly classified as 'hyperdiploid' when they are really 'hypodiploid'. The clinical significance of such an incorrect classification is unknown but in most instances is likely to be minor.

Two common problems in assessing DNA ploidy status are the difficulties in recognizing either small DNA aneuploid stem lines or, even worse, small tetraploid stem lines. These problems can be more easily dealt with if other flow cytometric parameters, which are freely available on most machines, such as 90° light scatter and forward light scatter (or Coulter volume) are used (see Fig 5.5) in conjunction with DNA content.

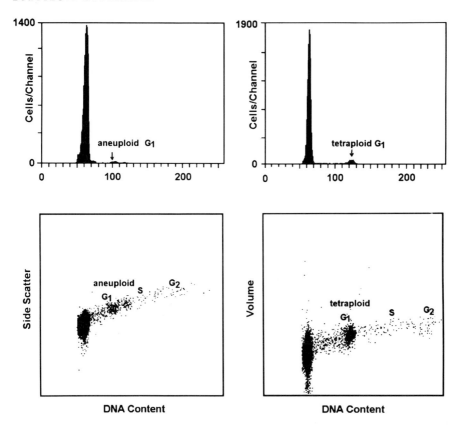

Fig. 5.5. The two plots on the left illustrate data from a brest carcinoma containing a tiny DNA aneuploid peak. On the histogram plot this peak could be missed but, by looking at the scatter plot below, a clear aneuploid population consisting of G_1, S and G_2/M cells can be seen. On the right-hand side are results for a tumour containing a very small tetraploid population. It would not be possible to distinguish this on the single-parameter DNA histogram from a diploid profile. However, tetraploid G_1, S and G_2/M cells can be seen on the scatter plot. These plots were produced on a FACS Analyser, an old mercury arc-lamp machine. Scatter measurements are generally still more informative on modern laser-based machines

In many published studies, recognition of tetraploidy is based on an arbitrary decision concerning the size of the peak in the 4C region of the histogram. Some authors assume a profile to exhibit tetraploidy if more than 10% or 20% of cells fall in this region. Based on our experience, this type of classification may lead to serious errors. By using other parameters such as forward/side scatter, small tetraploid populations may be clearly recognized by reference to the presence or absence of tetraploid S and G_2/M cells (Fig. 5.5).

S-phase fraction

The calculation of SPF is even more fraught with problems. For the estimation of SPF, the method of Baisch *et al.* can be used for DNA diploid histograms and a modification of this method for DNA aneuploid histograms (Camplejohn *et al.*, 1989; Camplejohn, 1992). These methods involve fitting a rectangular area to represent the S phase and calculating the number of cells within this area, as shown in Fig. 5.3. This method of calculation is recommended for a variety of reasons. Firstly, it is simple and requires only a hand-held calculator to perform. Thus, it can be applied anywhere and does not require any specific computer hardware or software. Secondly, it gives good agreement with BrdUrd labelling index in a series of cell lines.

It cannot necessarily be stated that this is the best method of analysis but at least its basis is understood. Many commercial computer programs will estimate ploidy and SPF from DNA histograms. Some of these programs are seriously flawed and the user may not understand the algorithm on which the calculations are based. Commercial manufacturers of flow cytometers generally supply software to analyse DNA histograms. In addition, such software can be bought from specific software houses to run on IBM-compatible computers. In either case, it is important to understand the assumptions involved in the estimation of SPF within a particular program.

One major drawback to DNA flow cytometry as a means of providing a crude assessment of proliferative activity is that a significant minority of histograms are uninterpretable. If fixation of tissue is appropriate, successful DNA profiles may be obtained from over 90% of tissue blocks. However, even in such a favourable case, SPF may only be calculable for 70–75% of total cases. This results from DNA histograms with (i) multiple aneuploid peaks, (ii) small aneuploid peaks, or (ii) peaks very close together. These problems are illustrated in Fig. 5.6.

SUMMARY OF METHODOLOGY

There are clearly many technical differences between the laboratories around the world in which DNA flow cytometry is performed. Thus, it is not possible to assume, for example, that an SPF of 5% is equivalent in all laboratories. However, as will be discussed shortly, trends shown by different laboratories may be similar even if the actual values achieved are not. Our own experience, also, is that if one does use a similar technique, then similar results are achieved on different sites.

CLINICAL VALUE OF DNA FLOW CYTOMETRY

CRITERIA FOR USEFUL CLINICAL STUDIES

Virtually all clinical DNA flow cytometric studies are performed in situations where routine histopathological techniques are also applied. Standard his-

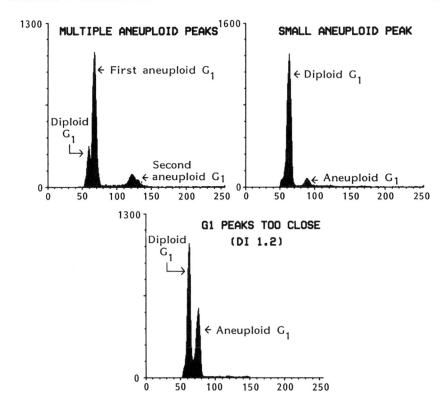

Fig. 5.6. This figure illustrates three types of DNA profile, from which we would not attempt to estimate SPF. Top left: this histogram has two DNA aneuploid stem lines (DI 1.1 and 2.0) and is too complex to allow our simple method of analysis. There are computer programs which will attempt to fit such data. Top right: this plot demonstrates a small aneuploid population; we do not calculate SPF if the total aneuploid population represents < 10% of all cells. One could reduce this threshold by running a very large number of total cells (5–10×10^4), although this would be more time consuming and would be more demanding in terms of data storage space. Lower: in this histogram the DNA aneuploid G_1 peak is close to the diploid peak and the two S phases largely overlap each other. It is possible to estimate a joint diploid and aneuploid S phase but the validity of this approach would need to be established

topathological techniques are well established and it seems unlikely that flow cytometry will to any extent replace such methods. Thus, if flow cytometry is to be performed, as well as routine histopathology, it must add something to it. Further, for routine clinical application, the additional information gleaned from flow cytometry should be of use in patient management. It should help in making specific choices as to how best to manage individual patients. In recent years, our clinical DNA flow cytometric work has been concentrated on areas where the above criteria might apply.

THE APPLICATION OF DNA FLOW CYTOMETRY IN PROGNOSIS

Much early effort was expended in testing the value of DNA flow cytometric measurements as an adjunct to the diagnosis of malignant disease. In general, such DNA measurements have not been of much value in the discrimination of benign from malignant conditions. There are some exceptions, for example in bladder cancer, but even here the technique is not widely applied. Certainly, measurements of DNA ploidy alone are of little value in diagnosing malignancy.

However, the literature on clinical studies of DNA flow cytometry in which the aim has been to compare DNA results with other known prognostic factors and/or to relate the flow results directly to patients' survival is vast. It is therefore impossible, in the space available here, to provide any sort of detailed review of the literature. But a number of review articles have been published, which would give an interested reader a starting point in finding data on a wide variety of tumour types (Macartney and Camplejohn, 1994; Merkel and McGuire, 1990; Raber and Barlogie, 1990). In this short section, two major types of cancer, namely non-Hodgkin's lymphoma (NHL) and breast cancer, will be considered as examples of the type of clinical data available. Both types of disease have been well studied to the extent that even restricting discussion to them will require a fairly superficial overview. But it has to be said that in both NHL and breast cancer clinically interesting results have been achieved. A brief comment will then be made on other major types of malignant disease.

Non-Hodgkin's lymphoma

DNA aneuploidy is more common in high-grade NHL but the presence of DNA aneuploid cells has little prognostic significance. Three studies in high-grade NHL found that DNA aneuploidy had an adverse effect on survival and one study found the opposite (Macartney and Camplejohn, 1990). In none of the above studies was an effect demonstrable after multivariate analysis.

Because of inconsistent flow cytometric methods, inconsistent pathological grading and uneven statistical analysis, it is difficult to compare the published reports of proliferative activity in NHL (Macartney and Camplejohn, 1990). In general, proliferative activity does correlate with histological grade. There is, however, in all published studies, substantial heterogeneity in proliferative activity within grades or histological subtypes of NHL, as well as between them. This limits the value of proliferative activity as a means of classifying or grading NHL. However, this very heterogeneity, within supposedly homogeneous histological subtypes of NHL, may make this measurement of prognostic value within a given subtype. As discussed earlier, any prognostic information must be additional to that obtained from standard histopathology, otherwise DNA flow cytometry is little more than an expensive exercise in tumour grading.

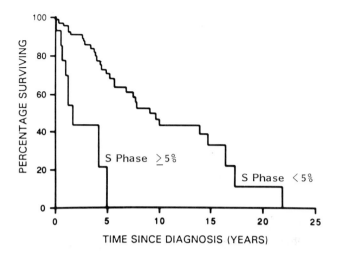

Fig. 5.7. Survival curves for follicular lymphomas with SPF above (*n* = 16) and below (*n* = 67) 5%. SPF retained its significance when treated as a continuous variable. Data from Macartney *et al.* (1991)

There are four studies which report a worse prognosis for low-grade follicular NHL with high proliferative activity (SPF > 5%), either in terms of overall survival or increased probability of high-grade transformation. Data from one of these studies (Macartney *et al.*, 1991) are given in Fig. 5.7. The two most recent studies (Rehn *et al.*, 1990; Macartney *et al.*, 1991) both show that B symptoms and high proliferative activity were the strongest predictors of survival in multivariate analyses. Thus, low-grade follicular NHL may be the sort of malignancy where DNA flow cytometry can aid patient management. There seems to be a case for treating follicular NHL that have high proliferative activity with more aggressive therapeutic regimes.

The situation in high-grade lymphoma is more complex (see Macartney and Camplejohn, 1990) but there does seem to be an adverse effect of high proliferative activity on survival, at least in the short term.

Breast cancer

A large number of flow cytometric studies have been performed on this disease. Many of the earlier studies were restricted to DNA ploidy and many of them looked only at the association between flow cytometric parameters and other pre-treatment variables, not at survival. To summarize the findings of these studies (see Merkel and McGuire, 1990, for bibliography), DNA aneuploidy is more common in high-grade tumours in virtually all published studies. The correlation of ploidy with other clinicopathological variables is less certain. Some groups report an association between the frequency of occurrence of DNA aneuploidy and tumour size, while other groups did not find this, and a similar story is to be found with nodal status and DNA ploidy. The relationship between steroid receptor status and DNA ploidy is also uncertain. Merkel and McGuire (1990) found 11 studies involving 2553 patients in which DNA aneuploidy occurred more commonly in oestrogen receptor-negative tumours and six studies involving 752 patients in which no relationship or a non-significant trend was found. The impression from reviewing the literature is that many of these parameters may be interrelated but that the correlations are weak.

As regards SPF, there is little evidence that it correlates with tumour size or nodal involvement, but many studies do find a correlation with steroid receptor status. SPF does correlate, usually strongly, with tumour grade in studies in which this association was investigated.

In terms of predicting survival, two distinct questions should be kept in mind. Firstly, do DNA ploidy and/or SPF predict survival in meaningful groups of breast cancer patients? Secondly, is any such predictive power independent of other known clinicopathological factors? As regards DNA ploidy, there is some disagreement for both node-negative and node-positive patients in answering both of these questions. There is now a vast literature in this field but overall it appears that most studies find, at least on univariate analysis, an association between DNA ploidy and survival, but that if a multivariate analysis which includes tumour grade is performed this predictive power is in most cases lost.

With regard to SPF, there is somewhat more agreement but even here there are studies with disparate results. In both node-negative and node-positive patients, tumour SPF was found in virtually all published studies to correlate usually strongly with survival (see Merkel and McGuire, 1990, for a summary of earlier studies; and Camplejohn, 1993, for a more recent comment). There are, however, disagreements as to whether SPF has independent prognostic significance, particularly, as with DNA ploidy, if tumour grade is included in the multivariate analysis. Several studies have found SPF (or SPF + G_2/M) to be of independent value in a multivariate analysis including grade. A number of other studies have reported it to be of independent prognostic value when other clinicopathological parameters such as steroid receptor status, tumour size, age, etc., were included but these studies did not include tumour grade. In general, the impression is that SPF might overall be a stronger prognostic indicator than DNA ploidy in breast cancer.

As an example of studies focused, as set out earlier in the criteria for clinical studies, on a specific clinical question, there follows a brief description of an investigation into node-negative breast cancer. A clinical problem with node-negative disease is that, although survival overall is better for patients with no nodes involved than for those with involved nodes, there is a subgroup of node-negative patients who do poorly. It would not generally be acceptable to give all node-negative patients toxic adjuvant therapy but it

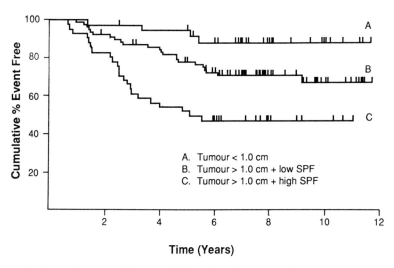

Fig. 5.8. Relapse-free survival curves for lymph node negative breast cancer patients: (A) tumours < 1.0 cm; (B) tumours > 1.0 cm + low SPF; (C) tumours > 1.0 cm + high SPF. Data from O'Reilly *et al.* (1990)

might be worth giving such treatment to suitable patients with a poor prognosis. How to identify such patients is therefore a clearly relevant clinical question. To see whether DNA flow cytometry might be a useful means of doing this, O'Reilly *et al.* (1990) suggested that by combining tumour size and SPF, a useful discrimination of a poor-prognosis subgroup of node-negative patients could be made (see Fig. 5.8).

Tumours from other sites

Having vastly simplified the above overviews of DNA flow cytometry in breast carcinoma and NHL, it is difficult to make even briefer comments on other major types of tumour. Nevertheless, an attempt will be made to give my impression of what are often large sets of publications. For more information, reference can be made to Macartney and Camplejohn (1994). There is no doubt that DNA aneuploidy and high SPF are in general associated with a poorer outcome. The crucial questions are, firstly, whether these two parameters are *independent* guides to prognosis and, secondly, if they are, do they constitute the best way of obtaining this information? On balance, there seems good evidence that DNA flow cytometry could have a role in the assessment of ovarian and endometrial cancer as well as breast cancer and NHL. It is possible, though less convincing, that the same could be true in bladder and prostatic cancer. The case for a clinical role for DNA flow cytometry in cancer of the lung and gastrointestinal tract seems weaker.

MULTI-PARAMETRIC FLOW CYTOMETRY

DNA AND NUCLEAR PROTEINS

All flow cytometry should be multi-parametric in the sense of making use of parameters freely available; these include light scatter (forward and $90°$; see Fig. 5.5) and pulse/height of the DNA signal. However, in this case, the other parameters are being used to improve the quality of the parameter of interest, namely nuclear DNA content. By using the additional parameters, debris and clumps can be 'gated out' of the analysis and it may be possible to distinguish between types of cell or nucleus (e.g. stromal versus tumour nuclei).

In contrast to this, there are studies in which two or more fluorescent compounds are used to measure or identify two distinct molecular species within the nucleus. In the research laboratory, flow cytometric measurement of antibody staining combined with DNA content is a powerful technique. For example this method has been used to define the cell cycle-related staining patterns of a variety of nuclear proliferation-related proteins. These include Ki-67 antigen, PCNA, p53, c-myc protein and many others. For example, flow cytometric studies have recently provided interesting data on the role of p53 in controlling entry of cells with damaged DNA (e.g. by ultraviolet or γ-radiation) into the S phase of the cell cycle.

An example of data from a multi-parametric study of PC10, a monoclonal antibody directed against proliferating cell nuclear antigen (PCNA), is illustrated in Fig. 5.9. PC10 has been used immunohistochemically on clinical material as a proliferative marker; it has the major advantage of working on paraffin-embedded sections. However, interpretation of PC10 staining on clinical material is complex and one reason for this has been disagreement about what cells it actually labels. Does it label only S phase cells or cells in all phases of the cycle? It has been shown using multi-parametric flow cytometry that this can depend on the staining protocol used (Fig. 5.9). The ability to obtain fresh cells by needle aspiration opens the possibility of performing conveniently such multi-parametric measurements on clinical material. A number of studies have been performed involving measurement of DNA content as well as markers such as Ki-67, PCNA and p53 in clinical material. It is too early to assess how useful such measurements may be in routine clinical use but preliminary claims for p53 and Ki-67, for example, have been made.

In passing, it is worth noting that some nuclear antigens may be preserved in nuclei extracted from paraffin-embedded sections. In such cases (e.g. p53 and c-myc), it may be possible to measure the level of the antigen and DNA content using the same disaggregation technique as illustrated in Fig. 5.4. Indeed, it has recently been demonstrated that the antigen recognized by a novel proliferation-related monoclonal antibody called Ki-S1 is retained in nuclei from paraffin sections (Camplejohn *et al.*, 1993). However, caution is required in the interpretation of such results, as redistribution and/or partial

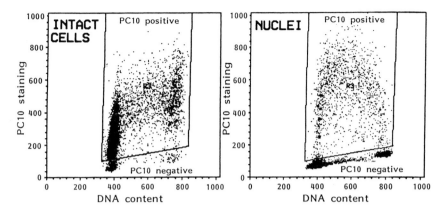

Fig. 5.9. The left-hand panel depicts the bivariate plot of DNA content versus PC10 staining (PC10 is a monoclonal antibody directed against PCNA) on intact cells of a human breast carcinoma cell line. Virtually all cells are labelled, with the intensity of staining showing a modest increase as cells pass through S phase and into G_2/M. The panel on the right shows staining with PC10 that is restricted to S phase cells. These results were obtained by treating cells with detergent prior to fixation. This process removes about 75% of PCNA, leaving only the PCNA bound tightly to replicating DNA

degradation of specific proteins can occur during fixation and processing. It should be stressed that multi-parametric fluorescence measurements such as those discussed above can be performed perfectly well on simple-to-use bench-top machines.

DNA AND BrdUrd

A specific example of a multi-parametric technique which is already being applied clinically, albeit in a particular specialized project, concerns the measurement of DNA content plus incorporated bromodeoxyuridine (BrdUrd). All simple indices of proliferative activity, such as [³H]thymidine labelling index, mitotic index, S phase fraction and Ki-67 labelling index give only a static measure. They indicate the number of cells in part or all of the cell cycle, but they fail to give any clue as to the *rate* of cell proliferation. Complex techniques, such as the labelled mitosis method, exist to obtain real kinetic information in experimental systems but they are inappropriate and impractical for use on patients. Recently, a technique involving injection of the thymidine analogue BrdUrd into patients 4–6 h prior to biopsy has been developed. Tissue from the biopsy is fixed in alcohol and processed with pepsin to yield free nuclei, which are then labelled with an anti-BrdUrd antibody and stained for DNA content. This nuclear suspension is then subjected to multi-parametric flow cytometry. By waiting 4–6 h between BrdUrd injection and biopsy, information can be gleaned about both the numbers of S phase cells and

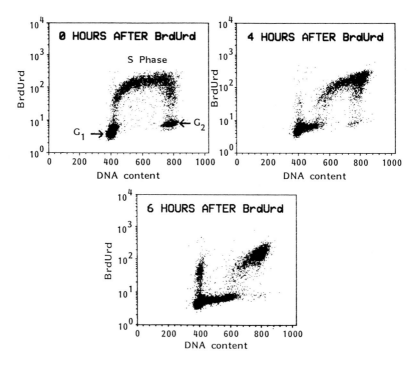

Fig. 5.10. These plots illustrate data obtained from the same cell line as illustrated in Fig. 5.9, labelled this time with an anti-BrdUrd monoclonal antibody. Top left: cells fixed immediately after incubation with BrdUrd. All S phase cells are labelled, G_1 and G_2/M cells are negative. Top right: cells were left for 4 h after BrdUrd incubation prior to fixation. Cells in the first 30–40% of S phase are BrdUrd negative, whilst virtually all G_2/M cells are positive due to passage of labelled cells from S into G_2/M. Lower: by 6 h most cells in S phase are unlabelled, all G_2/M cells are labelled and a clear cohort of labelled cells have reached G_1. From the rate of movement around the cell cycle, the potential doubling time (T_{pot}) can be calculated. By making various assumptions T_{pot} can be calculated from a single sample, thus making the technique suitable for clinical application. From Carter *et al.* (1990) by copyright permission of Wiley–Liss

their rate of passage round the cycle (Fig. 5.10). At the Mount Vernon Hospital near London, and at other sites, this technique is being applied to patients involved in a clinical trial of accelerated hyperfractionated radiotherapy (CHART) in which three doses of radiation are given per day for a period of 12 days (Wilson *et al.*, 1988). The aim of CHART is to counteract the effects of tumour repopulation, which can occur during a conventional radiotherapeutic schedule, spread typically over a 6-week period with gaps in treatment at weekends. A hypothesis has been proposed that CHART should benefit most those patients with rapidly growing tumours. BrdUrd labelling and flow cytometry are being used to test this hypothesis. Clearly BrdUrd labelling will not be applied routinely in histopathology to give an index of proliferative activity. It may, however, have application in a variety of circumstances, such as CHART, in which real kinetic data are required.

THE USE OF FLOW CYTOMETRY IN 'REAL' MOLECULAR STUDIES

Compared with many of today's molecular biological techniques, measures such as total cellular DNA content obtained by flow cytometry seem rather crude. This is not to say that they may not be useful. It is beyond the scope of this chapter to discuss the application of flow cytometry to 'molecular' studies at a finer level of individual chromosomes and even genes. However, one probable major future clinical application of flow cytometry is to provide sorted human chromosomes (e.g. 300 copies of a particular chromosome) from individual patients, which can then be used as the starting material for a variety of investigations. Clearly, a cell sorter is required for such investigations and a machine capable of employing simultaneously two lasers generating different wavelengths of light is generally required. Methods are available to generate suspensions of individual chromosomes, which by dual staining with two DNA dyes can be distinguished one from another, to produce a flow cytometric karyotype (Fig. 5.11). Most chromosomes can be recognized as

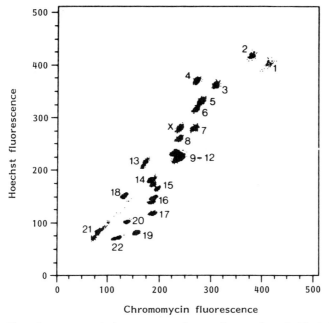

Fig. 5.11. Flow karyotype of chromosomes from a human lymphoblastoid cell line derived from a female (thus no Y chromosome). Chromosomes were double stained with two DNA dyes, namely Chromomycin A3 and Hoechst 33258. As shown, all chromosomes can be resolved into single peaks except for chromosomes 9–12, which appear as a single fused group. Data from Carter *et al.* (1990), by permission of Wiley–Liss

distinct entities in this way and sorted into tubes. PCR technology can then be used to amplify the DNA. This material can then be subjected to a range of molecular techniques such as fluorescence *in situ* hybridization (FISH). Such methods are useful in routine clinical cytogenetics to detect insertions, deletions and translocations. These defects can be involved in a wide range of pathological conditions. A series of five such clinical cases is discussed by Carter *et al.* (1992).

SUMMARY

DNA flow cytometry certainly seems to be of clinical value in a variety of tumours, including NHL and breast cancer. Whether it is the best or most cost-effective means of assessing proliferative activity in clinical material remains to be established. Histopathologists may prefer a histochemical method of obtaining a crude proliferative index by using proliferation-related antibodies on tissue sections (see Camplejohn, 1993). DNA flow cytometry may nevertheless have a role, for example, where ploidy as well as SPF is desired. Certainly, DNA flow cytometry is in routine use, for example, in breast cancer. In the USA the majority of breast cancer patients have DNA flow cytometry performed as part of their routine work-up (McGuire, personal communication). A single laboratory in California runs between 30 000 and 40 000 breast tumour samples per year.

Ideally, it would be best to standardize pathological classification of disease types, the flow cytometric and staining techniques and the methods of data analysis as outlined by Macartney and Camplejohn (1990) for NHL. However, in practice, this is hard to achieve. This does not prevent the application of DNA flow cytometric data to patient management but it does require caution in comparing results from different centres, unless they use similar methods. It would be hoped that the general trend of results would be similar in different centres but the precise values of, for example, SPF may not be. There are certainly some steps towards standardization which could be taken, notwithstanding the existing difficulties. For example, all current and future studies should look at the clinical value of flow cytometrically derived data in a meaningful multivariate analysis, which includes, where possible, all known relevant clinicopathological parameters. In this way, the significance of the results obtained should be far easier to assess.

REFERENCES

Camplejohn RS (1992) Flow cytometry in clinical pathology. In: Herrington CS and McGee JOD (eds), *Diagnostic Molecular Pathology: A Practical Approach*. Oxford: IRL Press.

Camplejohn RS (1993) A role for proliferative measurements in clinical oncology? *Ann. Onco.*, **4**, 184–186.

Camplejohn RS, Macartney JC and Morris RW (1989) Measurement of S-phase fractions in lymphoid tissue comparing fresh versus paraffin-embedded tissue and 4',6'-diamidino-2 phenolindole dihydrochoride versus propidium iodide staining. *Cytometry*, **10**, 410–416.

Camplejohn RS, Brock A, Barnes DM *et al.* (1993) Ki-S1, a novel proliferative marker: flow cytometric assessment of staining in human breast carcinoma cells. *Br. J. Cancer*, **67**, 657–662.

Carter, NP *et al.* (1990) Study of X-chromosome abnormality in XX males. *Cytometry*, **11**, 202–207.

Carter, NP, Ferguson-Smith MA, Perryman MT *et al.* (1992) Reverse chromosome painting: a method for the rapid analysis of aberrant chromosomes in clinical cytogenetics. *J. Med. Genet.*, **29**, 299–307.

Hedley DW, Friedlander ML, Taylor IW, Rugg CA and Musgrove EA (1983) Method for analysis of cellular DNA content of paraffin-embedded pathological material using flow cytometry. *J. Histochem. Cytochem.*, **31**, 1333–1335.

Macartney JC and Camplejohn RS (1990) DNA flow cytometry of non-Hodgkin's lymphomas. *Eur. J. Cancer*, **26**, 635–637.

Macartney JC and Camplejohn RS (1994) DNA ploidy: flow cytometric aspects. In: Hamilton PW and Allen DC (eds), *Quantitative Clinical Pathology*. Oxford: Blackwell Scientific Publications (in press).

Macartney JC, Camplejohn RS, Morris R *et al.* (1991) DNA flow cytometry of follicular non-Hodgkin's lymphoma. *J. Clin. Pathol.*, **44**, 215–218.

Merkel DE and McGuire WL (1990) Ploidy, proliferative activity and prognosis: DNA flow cytometry of solid tumors. *Cancer*, **65**, 1194–1205.

O'Reilly SM, Camplejohn RS, Barnes DM *et al.* (1990) Node negative breast cancer: prognostic subgroups defined by tumour size and flow cytometry. *J. Clin. Oncol.*, **8**, 2040–2046.

Raber MN and Barlogie B (1990) DNA flow cytometry of human solid tumors. In: Melamed MR, Lindmo T and Mendelsohn ML (eds), *Flow Cytometry and Sorting*, 2nd edn. pp 745–754 New York: Wiley–Liss.

Rehn S, Glimelius B, Strang P, Sundstrom C and Tribukait B (1990) Prognostic significance of flow cytometry studies in B-cell non-Hodgkin lymphoma. *Hematol. Oncol.*, **8**, 1–12.

Vindelov LL and Christensen IJ (1990) A review of techniques and results obtained in one laboratory by an integrated system of methods designed for routine clinical flow cytometric DNA analysis. *Cytometry*, **11**, 753–770.

Wilson GD, McNally NJ, Dische S *et al.* (1988) Measurement of cell kinetics in human tumours *in vivo* using bromodeoxyuridine incorporation and flow cytometry. *Br. J. Cancer*, **58**, 423–431.

FURTHER READING

Melamed MR, Lindmo T and Mendelsohn ML (eds) (1990) *Flow Cytometry and Sorting*, 2nd edn. New York: Wiley–Liss.

Ormerod MG (1990) *Flow Cytometry: A Practical Approach*. Oxford: IRL Press.

Wilson JV (1991) *Introduction to Flow Cytometry*. Cambridge: Cambridge University Press.

6 Molecular and Immunological Aspects of Cell Proliferation

J. CROCKER

THE CELL CYCLE AND ITS IMPORTANCE IN CLINICAL PATHOLOGY

In recent years, it has become increasingly apparent to histopathologists that a simple description of the morphological appearance of a section of tissue cannot give sufficient idea of its behavioural status. This is most unfortunate, since the accurate diagnosis of a specimen is linked to effective treatment of the patient and further management. It has also become clear to us that in this context the description of individual tissues (notably cancers), in terms of their histogenetic type, is insufficient. Modern diagnosis of many biopsy specimens depends upon various changes which are directly or indirectly related to cell replication. This is particularly true in the field of cancer, where cell division is known to be disordered. In general, such assessments are largely subjective and it is therefore highly desirable that more objective techniques become available. An example of a 'problem tumour' lies in leiomyosarcoma of the stomach; this is a neoplasm encountered by surgeons on occasion and which has a very variable prognosis. This latter is determined, presumably, by various factors in the biology of each case; however, variables as divergent as tumour mass and mitotic counts have been promulgated as indicators of behaviour. Nonetheless, it would seem to be logical that the level of cell proliferation in a tumour might be one of the more important features governing its future behaviour. It must never escape our understanding that factors including cell motility and adhesion, destruction of the basement membrane by proteinases and escape from the immune system may be at least as important as cell proliferation in the natural history of a malignant tumour. None of us should therefore be in the least surprised if measurement of the proliferative ability of a specimen or series of specimens of cancer does not relate well to progression or clinical parameters of survival. Notwithstanding this caveat, the assessment of cell proliferation has proved to be most useful in many contexts in histopathology and it is timely that the immunological means available for this are described later. Other methods can afford information regarding cell proliferation; some of these are described in Chapters 5 and 7.

Molecular Biology in Histopathology. Edited by J. Crocker
© 1994 by John Wiley & Sons Ltd

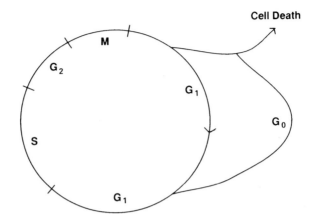

Fig. 6.1. Schematic diagram of the stages of the cell cycle. Cells can exit from this, either permanently or temporarily. $G_{0,1,2} = Gap_{0,1,2}$; S = synthesis; M = mitosis

The molecular events in the cell cycle are increasingly understood and form a basis for our investigations into pathological tissue; accordingly, these are outlined below.

THE EVENTS AND STAGES OF THE CELL CYCLE

Since the 1950s, it has become clear that the synthesis and replication of DNA through cell division is not a continuous phenomenon. Pulse-labelling with radiolabelled precursors showed that there were 'gaps' in the process and that these were of variable duration in different cell types, as were the other phases of what has become known as the 'cell cycle' (Fig. 6.1). The highly complex series of molecular control mechanisms involved in this system are outlined below.

THE CELL CYCLE ITSELF

The *raison d'être* of the cell cycle is the entry into and completion of mitosis, leading to cell proliferation and increase in numbers. In general, once a cell is committed to mitosis, the process is irreversible, although it is possible at certain points to exit from the process. The enumeration of mitotic profiles in a tissue section can give an estimate of the level of proliferation, although even this apparently simple technique is fraught with difficulties. Some types of cells in normal tissues have greater or lesser frequencies of entry into the cycle. For example, lymph nodal follicular centroblasts have a high frequency of cycling, whereas the mantle zone lymphocytes do so with a low frequency. Indeed, some cells, such as mature neurones, do not divide. The picture is complex in cancer cells, where a great range of cell turnover rates is observed; nonetheless, it is the case that in general malignancies of high aggressiveness have more cells in cycle at a given time than do low-grade tumours.

The cell cycle can be regarded as having different phases, which can be related to molecular changes and events. The stages comprise: G_1 ('gap 1'), which spans between mitosis and DNA synthesis; S ('synthesis') when DNA is synthesized; G_2 ('gap 2'), which is the period between the end of DNA synthesis and the commencement of mitosis, and M, mitosis itself. In most organisms, the greatest variation of cycle phase duration in proliferating cells is in G_1 (as opposed to $S + G_2 + M$), implying that it is in this stage of the cycle that most regulation occurs. The critical nature of G_1 is further underlined by the numerous growth factors and nutritional agents required for its successful completion and by the fact that permanent or reversible exit from the cycle occurs at this stage. However, the picture is complicated by the existence of the so-called restriction point or 'START', which lies in G_1; once a cell has passed this point, entry into S phase is inevitable even if growth and nutritional factors are absent (Pardee, 1974). Interestingly, it has been shown recently that p34^{cdc2}, the protein product of the gene *cdc2* (see below), is necessary at the START point; this came as something of a surprise when discovered, since it was already known that passage from G_2 to mitosis was dependent upon a protein kinase composed principally of p34^{cdc2} and cyclins. The latter are described later. If cell-free extracts of frog oocytes are prepared, and chromatin or DNA added, then DNA synthesis continues, even in the absence of p34^{cdc2}; in the latter case, however, DNA synthesis cannot be *initiated*. It may be that this initiating step is at START. It has also been shown in T lymphocytes that the initiation of DNA synthesis can be blocked with antisense oligonucleotides to *cdc2*. Other genes, including CLN-1 and *cdc13*, have also been implicated in the regulation of START, at least in yeasts. Cyclin (of the A variety) is involved here by activating p34^{cdc2} so that the cell passes from G_1 to S phase and specific 'G_1 cyclins' are now being sought in a variety of organisms; indeed, a cyclin C, three cyclins D and a cyclin E have now been found in mammals and may act as true G_1 cyclins. Much of our current understanding of the control of cell division derives from studies of frog oocytes, which arrest in meiosis (the first division) and mature irreversibly to ova. This maturation is, perhaps not surprisingly, dependent upon protein synthesis and is induced by the presence of progesterone. Remarkably, it was shown that this maturation could be induced very rapidly, in the absence of progesterone, by the introduction of cytoplasm from mature ova, suggesting that the latter contained a substance (or substances) which was responsible for maturation. This was named 'maturation promoting factor' or MPF. Further, it transpired that MPF could induce maturation in the absence of protein synthesis, indicating that an inactive precursor ('pre-MPF') was present in the oocyte prior to stimulation. As the ovum begins to divide, levels of MPF rise with mitosis. This maturation phenomenon has been shown to be true in a very wide range of organisms (including humans), since cytoplasm from these organisms can induce frog oocyte maturation; indeed, the factor has generally been renamed *mitosis promoting factor*.

It was subsequently discovered that a cell-free system could be designed in which the chemistry of MPF could be more conveniently examined, and when attempts were made to purify MPF a 32 kDa protein was found which was closely related to a cell cycle-related protein, $p34^{cdc2}$, in yeasts (Lohka *et al.*, 1988). The gene responsible for the production of this protein is *cdc2* and its mutant variants in certain yeasts can lead to shortening of G_2, thus leading to earlier mitosis than usual, this in turn resulting in smaller dividing cells (the mutant phenotype being designated *wee*, accordingly!). $p34^{cdc2}$ appears to be highly evolutionarily conserved, and the human homologue is now known to exist.

Further analysis has led to the conclusion that MPF is largely equivalent to a histone H1 kinase, whose function is independent of cyclic nucleotides, calcium and diacyl glycerol. The kinase exhibits increased phosphorylation during mitosis and when introduced into cells is accompanied by increase in phosphorylation and early entry into division. Apart from histone H1, the kinase is also involved in the dissolution of lamins, which lie at the nuclear envelope and disappear as mitosis commences. Indeed, it now appears that the onset of mitosis probably involves a series of kinases, comprising a sort of 'cascade', the complexities of which are beginning to unfold.

CYCLINS

Cyclins have been recognized for over a decade and are proteins whose levels fluctuate greatly in marine invertebrate ova after fertilization has occurred and which accumulate during the cell cycle and are rapidly destroyed when mitosis occurs. Two types of molecule, cyclins A and B, are now recognized and these have both been described in humans as well as other organisms, cyclin A being destroyed rather earlier than B at mitosis (Luca and Ruderman, 1989). There are also certain fairly subtle differences between the two cyclins, beyond the scope of this review. The main role of the cyclins is to activate $p34^{cdc2}$, to which they can be shown to bind (Fig. 6.2). Regulation of the level of MPF is now known to be dependent upon the phosphorylation of $p34^{cdc2}$, with increasing phosphorylation through the S and G_2 phases of the cell cycle, together with incorporation into a larger, 200 kDa complex. Concomitant phosphorylation of cyclin also occurs. Dephosphorylation, notably of the tyrosine (and threonine and serine) amino acids of $p34^{cdc2}$ is necessary for activation to histone H1 kinase, as can be demonstrated by means of metabolic blockage experiments. Furthermore, this dephosphorylation may be dependent upon calcium ions and calmodulin.

Matters are, however, even more complex. Thus, it is now evident, at least in yeast, *Drosophila* and humans, that $p34^{cdc2}$ is up-regulated by the *cdc25* gene

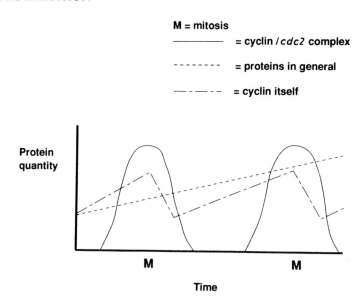

M = mitosis

———————— = cyclin / *cdc2* complex

- - - - - - - - = proteins in general

— - — - - = cyclin itself

Protein
quantity

M

M

Time

Fig. 6.2. The levels of cyclins and other proteins through the cell cycle

product ($p80^{cdc25}$), which is necessary for entry into mitosis, has a size of 80 kDa and is involved in the phosphorylation of tyrosine.

The destruction of cyclin is also of essential importance in the escape from mitosis. This proteolysis is catalysed by $p34^{cdc2}$ kinase and this in turn leads to the inactivation of $p34^{cdc2}$ itself. Only part of the structure of cyclin is necessary for mitosis-dependent destruction; this segment has recently been ascribed the title 'destruction box', although a further nearby section of the molecule is apparently also needed. Interestingly, there is now evidence that the mitotic spindle itself may also regulate cyclin breakdown.

G_1 AND CANCER CELLS

The events at G_1 can be seen to be very important in the control of cell turnover, as mentioned above, and it is at this point that certain external influences such as growth factors exert an influence on the cycle. For example, epidermal growth factor (EGF), which is a small peptide (53 amino acids), binds to its receptor in the cell plasma membrane, causing it to dimerize, resulting in autophosphorylation of its tyrosine kinase. This in turn leads to phosphorylation of tyrosine, serine and threonine residues, leading to the 'kinase cascade', as above. Conversely, some factors can act at G_1 and inhibit cell proliferation; these include tumour necrosis factor, interferons and transforming growth factor-β.

In cancer cells, proliferation is disorganized with incorrect timing and coordination, and defects can be detected at G_1 (Pardee, 1989). Indeed, cancer cells can grow *in vitro* in the absence of certain growth factors such as EGF and in concentrations of serum too low to support the growth of normal cells. Anchorage to substrate, needed for the growth of non-malignant cells, is also unnecessary when such cells are malignant. Interestingly, malignant cells often have more EGF receptors on their surfaces than normal, suggesting the availability of more potential for stimulation of proliferation. Cancer cells can, of course, also be in a quiescent state and it may be that this is different from simple G_0, since certain substances (including *myc* mRNA, a subunit of ribonucleotide reductase and 'protein p68') can be detected in arrested cancer cells but are absent from normal G_0 cells.

In conclusion, it will be apparent to the reader that cell cycle control is extremely complex and this should not surprise us. It must be recalled that although there is often great structural and functional evolutionary conservation of the molecules concerned, many of the studies in this field have been performed on much lower organisms, from yeasts to marine invertebrates. Great care must be employed in extrapolating to the human condition. The above is but an overview of this increasingly complex subject and readers requiring further information are encouraged to go to more detailed reviews and papers.

MITOSIS

The phases and mechanisms of mitosis (McIntosh and Koonce, 1989) are outlined in brief below:

(1) In prophase, the chromosomes condense, ready for transport; this is effected by means of microtubules, under the control of the centrosome, which divides in mitosis and increases and rearranges the microtubules. In addition, protein synthetic levels decline and RNA synthesis ceases.

(2) In prometaphase, one copy of each chromosome aligns with one end of the cell, by means of the 'metaphase plate'; this is organized by means of the centromere of the chromosome, which forms the kinetochore with associated, bound proteins. These structures drag the chromosomes into the correct orientation at the metaphase plate. As this is occurring, the nuclear membrane breaks down, as described above.

(3) In anaphase, the two copies of the chromosome separate and migrate to the ends of the cell.

(4) In telophase, the cell reverts to its interphase configuration, with decondensation of the chromosomes. After this, the cell itself divides, forming two daughter cells. This is an active metabolic event and requires the presence of an actomyosin-dependent contractile ring.

DNA REPLICATION

The fundamental importance of all the foregoing lies, of course, in the replication of the genetic material, DNA. Most of our understanding of this process has arisen from studies of frog oocytes, as before in cell-free systems, and of cells infected by the double-stranded DNA virus SV40. (In the latter, the host cell, human or primate, is the source of the relevant regulatory mechanisms). These investigations have yielded results that are comparable and are likely to relate to the position in humans.

Eukaryotic cells have > 10 000 points of replication ('replication forks') in their entire genome and these progress in both directions until each fork meets another. The timing and spatial organization of these processes are therefore critical and, inevitably, highly complex. Thus, for example, segments of DNA must not be omitted or reduplicated in the process of cell division. The advancement of the duplication at the fork is critical; thus, topoisomerase enzymes must unwind the approaching DNA and its 'super-helix' twists must also be unwound. The DNA is then duplicated, and this must be in forward and retrograde directions, since DNA polymerses are active only in the 5'–3' direction.

When SV40 replicates, it requires only two virus-encoded proteins and the process can be performed *in vitro*. The replication occurs in two phases, namely the formation of an initiation complex at the replication point and, secondly, the elongation process, where replication forks move forward on the DNA template. In the initiation phase, SV40-encoded 'large tumour antigen' (T antigen) binds at the replication point of origin. This then moves away, unwinding DNA as it does so, and as this occurs the resulting single-stranded DNA produced is stabilized by a host-encoded protein named 'RF-A'. Other host proteins needed in this process include DNA polymerase δ, proliferating cell nuclear protein (PCNA; see below) and topoisomerases. In the frog, there are many similarities in these processes, although they are far less clear than in the SV40 system. When frog oocytes are labelled with biotinylated dUTP, they can be submitted to flow cytometry and their newly synthesized DNA can be detected. By means of this method, it has been shown that 10^5 replication forks are formed when nuclear division occurs after fertilization. Where it is important that re-replication does not occur, control of this process is effected by so-called 'licensing factor', which apparently binds to potential sites of future DNA replication and which is destroyed after division, so that further replication cannot occur.

IMMUNOCYTOCHEMICAL MARKERS OF PROLIFERATING CELLS

GENERAL CONSIDERATIONS

It will be apparent from the foregoing that there are several, if not many, candidates for 'markers' of cells which are proliferating. These can be expressed throughout much or all of the replicative phases of the cell cycle or be highly restricted in expression to only a very limited portion of the cycle. Furthermore, some of the antigens defined by the available antibodies may be of unknown location and chemical conformation; in these cases, it has often been the case that their value has been found on an empirical basis only. Examples of this type of preparation lie in some of the autoantibodies to human nuclear and nucleolar molecules. Conversely, in more recent years, the chemical nature and ultrastructural localization have been determined for some of the epitopes or molecules binding to certain antibodies. Our greater understanding of the molecular events and control of the cell cycle should enable us to label specific phases of the cycle in cells or tissue sections, although it could be argued that for 'routine' diagnostic purposes such 'fine tuning' is at present unnecessary (or, at least, of unknown value!).

Certain general comments should be considered with regard to the practical aspects of the investigation of proliferative status by means of immunocytochemistry. Firstly, in general it is the nucleus which is labelled in these methods and it is then necessary to count the numbers of positively stained nuclei in relation to a certain number of overall cells per nuclei. The 'lazy' technique of enumerating cells per nuclei per high-power field is to be eschewed at all costs, since not only does the area of such a field vary considerably from microscope to microscope, but also the measurement becomes meaningless since cell size varies from specimen to specimen (Fig. 6.3). Accordingly, the score of 'events' per area tells us little of the frequency of such events per number of cells, which is of much greater biological significance. The minimum number of total cells to be included in the counting procedure should always be derived by the standard continuous mean method.

The next consideration is that of tissue heterogeneity. This is a particular problem in some neoplasms, where some of the cells may be residual from the original tissue or may be 'reactive' (for example, inflammatory or endothelial cells) (Fig. 6.4). A rather arbitrary correction factor has been proposed by some

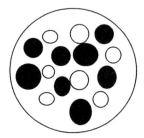 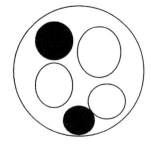

Fig. 6.3. The problems encountered on counting 'events per microscope field'. Depending on the overall cell size, counts for the positive structures (black) may seem more or less frequent relative to the negative structures (white). It is more satisfactory to express results as 'positive cells per n negative cells'

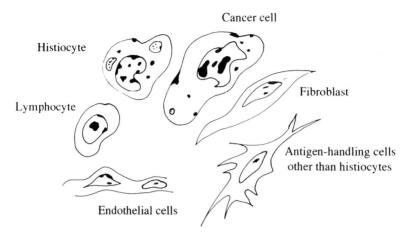

Fig. 6.4. Another problem encountered in enumerating positively stained cells in a malignant neoplasm. Many of the cells present may not be neoplastic and yet may express molecules leading them to be included in the score

workers to deal with this problem; however, the most stringent (and often the least practical) method is to use dual labelling of the tissue under investigation, with one or more other antibodies to highlight the cells of interest or to exclude others. This latter approach, although theoretically highly desirable, is very labour-intensive. The next practical point to be considered is that of the suitability of the tissue for study with a particular antibody. Unfortunately, many of the antibodies available for labelling proliferating cells will only succeed with frozen tissue, and routinely processed paraffin wax-embedded tissue is not suitable in this context. This can cause problems in two ways; firstly, paraffin wax-embedded archival material used in most studies in histopathology laboratories cannot be accessed; secondly, the morphology attained with frozen sections is never as clear as that available with a paraffin section. Thus, the cell types reacting with an antibody cannot be recognized with certainty on morphological grounds.

It has become apparent that the data derived from studies of, for example, neoplastic tissues using a particular antibody may not correlate well with those obtained from the use of other antibodies or methods. For example, staining with antibodies such as Ki-67 (see below) usually gives results in accord with those afforded by 'AgNOR' enumeration; conversely, other parameters may correlate poorly. The reasons for these observations are complex but probably lie in the fact that most of the antibodies in use label more than one phase of the cell cycle, unlike the uptake of thymidine or its analogues, which is restricted to the S phase. Furthermore, the half-life of some of the marker proteins may be relatively lengthy and, although they may not still be metabolically active, they may yet be detectable immunologically. Under ideal conditions, perhaps, any specimen to be evaluated for proliferation status should be subjected to more than one type of method. In reality, however, this cannot always be practical.

The 'classical' method for the assessment of proliferation in tissues or cells was to subject them to labelling with [³H]thymidine, which was incorporated into the DNA of S phase nuclei; however, the method required the availability of fresh tissue and involved cumbersome autoradiography, which was also time-consuming. Accordingly, thymidine labelling is rarely used today, having been largely supplanted by newer methodologies.

The following part of this chapter details the various antibodies available for the assessment of cellular proliferation in tissue sections. It should become apparent to the reader that although advances have been made in terms of application to paraffin sections, many problems still exist in the practical side of this field.

SPECIMEN HANDLING AND PREPARATION

For most immunohistochemical purposes tissue can be fixed in 10% formalin and processed in the conventional manner. To optimize fixation, it is recommended that the tissue is obtained fresh, as below, and that thin (2 mm) slices are placed in the formalin. However, for many markers of cell proliferation it is necessary to use frozen sections and it is therefore necessary either to stain freshly cut frozen sections or to maintain tissue at a low temperature until required for sectioning. The latter is the usual case, so it must be arranged with the surgeon that fresh tissue is available for collection from the operating theatres to enable appropriate handling. Slices of tissue, about 2–3 mm thick, should be dropped into liquid nitrogen, then stored in the same. In the author's experience, there is no need to make use of cooled isopentane to freeze tissues; this merely adds complexity to the process. Whether it is necessary to use frozen sections or whether paraffin sections are amenable, the immunohistochemical labelling process to be applied should be considered. Whenever possible, a chromogenic system which gives a permanent reaction product is preferable. Such a system is the avidin–biotin complex method, which has found widespread routine use. However, there are occasions when the alkaline phosphatase–anti-alkaline phosphatase technique can be invaluable, especially when high amplification of the 'signal' is necessary. This latter method may also be very helpful when double-staining (i.e. staining for two antigens or staining for one antigen plus *in situ* hybridization) is important. For details of the methodology, the reader should refer to one of the several practical texts in this field.

'TRADITIONAL' ANTIBODIES

Antibodies to transferrin receptor

Transferrin receptor (TfR) is a substance which, like its ligand transferrin, is essential to the life of many (if not most) living organisms. The uptake of iron is as necessary for life-forms as lowly as bacteria as it is to mammals, and the regulation of its usage is predictably quite highly controlled. The availability of iron binding is of great importance, since in the native state iron is in its ferric (Fe^{3+}) form and thus tends to be hydrolysed to insoluble $Fe(OH)_3$. Iron-binding molecules such as transferrin 'protect' iron from this process and each molecule of transferrin can bind up to two atoms of iron. The transferrin binds to TfR on the cell surface and the complex is then internalized via coated vesicles (Fig. 6.5). The highest rate of such iron uptake is observed in haemoglobin-synthesizing reticulocytes and in the placental trophoblast; in the former, each cell has up to 3.10^5 TfR sites and the entire sequence of iron/transferrin uptake, release of iron and return of apotransferrin to the exterior takes up to only 30 s. It is likely that all human cells possess TfR sites but at greatly divergent densities. Thus, to give an assessment of TfR positivity for a particular cell species is really a statement describing the ability of the antibody used to detect a certain amount of surface TfR. Monoclonal antibodies to TfR have been available for some years; two such antibodies are B3/25, which was raised against a human erythroid leukaemic cell line, and OKT9, which is also reactive with some T lymphocytes.

Biochemical analysis of TfR has been facilitated by the use of monoclonal antibodies against transferrin itself, to purify the transferrin/TfR complex or, more directly by means of OKT9, directed against TfR itself. It has been shown that the receptor is composed of two 90 kDa subunits, joined by disulphide linkage. The evidence is that TfR is a transmembrane molecule, with covalently bound fatty acid and phosphoryl serine residues. There is also a transmembrane 'tail' of 5 kDa size, as demonstrated by the protease treatment of microsomes. It also appears that each 90 kDa subunit of TfR binds to a single transferrin molecule. Recently, chromosome-mapping techniques have been applied to human–murine cell hybrids, to localize the gene encoding for TfR; these experiments have made use of the fact that the OKT9 antibody is species-specific for humans and that only the hybrid cells retaining the appropriate human chromosome will express TfR binding to this antibody. By this method, it has been shown that chromosome 3 is the location of the receptor gene, lying at the 3q26.2 position. (Interestingly, the location of the gene for transferrin itself lies close by, at 3q21. Furthermore, there is a protein gp97, associated with malignant melanoma cells and also capable of binding iron, which is encoded by a gene at 3q29. This suggests that the inheritance of these genes may be linked as a family.)

It was noted in the early 1980s that immunostaining for TfR might prove to be a useful method for the assessment of numbers of proliferating cells in a tissue sample. When fluorescent labelling with OKT9 was applied to leukaemic cells, with double tagging for DNA with mithramycin, followed by cell sorting, there was good correlation between immunolabelling and proliferation status. Induction of differentiation, with lowering of proliferation, was associated with a decrease in TfR expression. Furthermore, when the OKT9

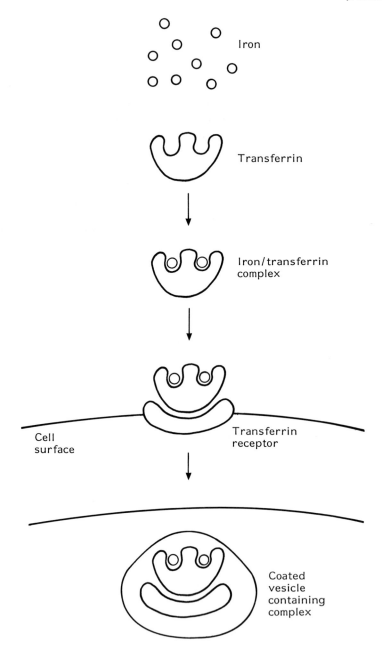

Fig. 6.5. Schematic diagram to illustrate the uptake of iron and transferrin by an active cell

antibody was applied to cell suspensions of cells from high- and low-grade lymphomas, there was a much greater level of labelling of the former (range 3–57%; mean 22.5%) than the latter (range < 1–22%; mean 2.5%) (Habeshaw *et al.*, 1983). Similar results have also been reported for carcinoma of the breast. Antibodies to TfR have also been applied to frozen sections of human tissue, although paraffin wax sections are not suitable. However, the most unfortunate limitation to the use of anti-TfR antibodies in the analysis of histological material, especially when it is neoplastic, is that they bind strongly to macrophages/histiocytes. This is an especially great problem, since most tumours, especially those of high malignancy, contain these cells. Although double immunostaining with anti-TfR and anti-macrophage antibodies could help in this context, it would be highly labour-intensive to perform this procedure and it must be admitted that the use of antibodies to TfR should be regarded now as being of academic rather than practical value in the field of histopathology.

Antibody BK 19.9

This antibody has found only limited use in the demonstration of proliferating cells, probably because it has not been fully characterized and because it can only be applied successfully to frozen section material (Fig. 6.6) (having been overtaken by the antibody Ki-67, as described below). BK 19.9 was raised against myeloid leukaemia cells and was shown by means of binding of anti-TfR and transferrin itself to co-cap (that is to say, to move to one pole of the cell) with TfR. It was therefore concluded that the antibody was directed against transferrin receptor. However, subsequent evaluation (Gatter *et al.*, 1983) demonstrated that this may not be the case. For example, hepatocytes are BK 19.9$^-$ but B3/25$^+$ and Kupffer cells are BK 19.9$^+$, although both cells are known to express TfR. Anomolous staining with BK 19.9 of pancreatic islets is also seen. Furthermore, expression of the BK 19.9 antigen precedes that of TfR during the entry of G_0 B lymphocytes into the cell cycle.

Antibodies to 5-bromodeoxyuridine

Antibodies to 5-bromodeoxyuridine (BrdUrd) and 5-iododeoxyuridine (IrdUrd) have been available for over a decade (Gratzner, 1982). As cells replicate, they incorporate BrdUrd into their DNA (since it is an analogue of thymidine) and this can then be detected by means of anti-BrdUrd antibody. Earlier studies with anti-BrdUrd made use of fluorescence-activation cell sorting (FACS) and, for example, in non-Hodgkin's lymphomas, gave results equivalent to those obtained with [^{3}H]thymidine uptake. Subsequently, the antibody was applied to cell suspensions from tumours, with more cells binding BrdUrd in high-malignancy specimens than in those with low

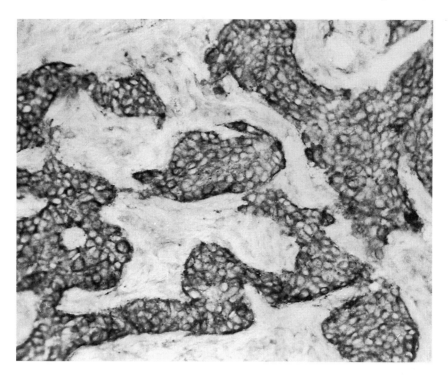

Fig. 6.6. A frozen section of a carcinoma of breast, showing extensive positive labelling with the antibody BK 19.9. Photomicrograph courtesy of Dr Rosemary Walker, Department of Pathology, University of Leicester

aggressiveness. An important advance lay in the discovery that fresh tissue slices can be incubated in a solution of BrdUrd, cut as frozen sections or processed to paraffin wax, then reacted with anti-BrdUrd antibody. To optimize the reaction, it is necessary to run the BrdUrd-binding reaction at a high oxygen tension, to 'drive' the metabolic uptake of this substrate. This can be done either by bubbling oxygen through the incubation medium or by increasing the pressure to at least 3 atmospheres. In addition, it is necessary to pre-treat the sections with hydrochloric acid to denature or unwind the DNA, to allow access of the antibody to the incorporated BrdUrd (Fig. 6.7). It must be emphasized that this is not a true 'paraffin method' since it requires the availability of fresh tissue. Furthermore, diffusion of the reaction mixture into the tissue sample is often limited and a good reaction may be seen only at the periphery of the section (Fig. 6.8).

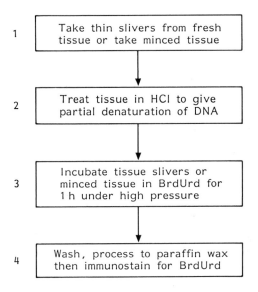

1 — Take thin slivers from fresh tissue or take minced tissue

2 — Treat tissue in HCl to give partial denaturation of DNA

3 — Incubate tissue slivers or minced tissue in BrdUrd for 1 h under high pressure

4 — Wash, process to paraffin wax then immunostain for BrdUrd

Fig. 6.7. Scheme for the labelling of fresh tissue slice with BrdUrd prior to immunostaining

Ki-67 antibody

In the early 1980s a novel antibody was described by a German group, having
been raised against the L 428 Hodgkin's disease cell line (Gerdes *et al.*, 1984). It
was found that Ki-67 reacted with cells known to be proliferating and could be
applied to tissue sections (Fig. 6.9), although a constraint was that paraffin
wax-embedded material was not amenable to staining with the antibody.
There have been many studies of wide-ranging tissues and these have
generally shown a good correlation between Ki-67[+] cell numbers and tumour
grade; furthermore, in some instances the Ki-67 score has been related closely
to survival or prognosis. The Ki-67 antigen is expressed by cells in all phases of
the cycle other than G_0 and early G_1 and is maximal in amount in G_2 and M
phases. In general, there is a good correlation between the Ki-67 index and
other measures of cell proliferation such as [³H]thymidine and BrdUrd uptake,
DNA flow cytometric analysis and interphase AgNOR scores. There have
been two recent significant advances in relation to Ki-67; firstly, the antigen has
been much better characterized than before and, secondly, an antibody with
the same reactivity has been produced which can be applied to paraffin
wax-embedded tissue sections. These developments are described below. It
has been shown that on western blotting of proliferating (but not resting) cells
there are two very large polypeptides, of 345 and 395 kDa sizes, which react
with Ki-67. When immunoscreening of human λT11 cDNA libraries was
performed, 10 Ki-67[+] clones were found, of which eight had identical DNA

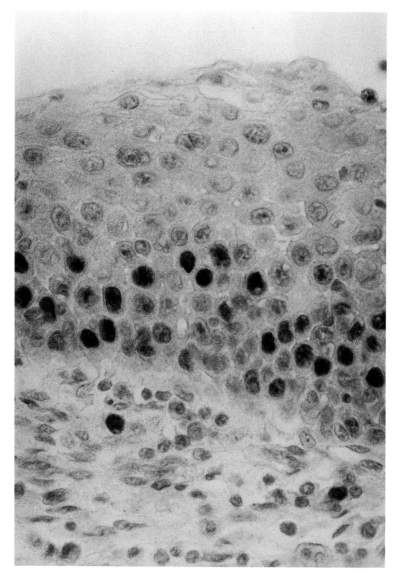

Fig. 6.8. Tonsillar pharyngeal squamous epithelium which has been incubated with BrdUrd then immunostained with anti-BrdUrd antibody. As would be expected, most of the nuclei in the suprabasal layer are positively stained. Preparation courtesy of Mr Paul Murray, University of Wolverhampton

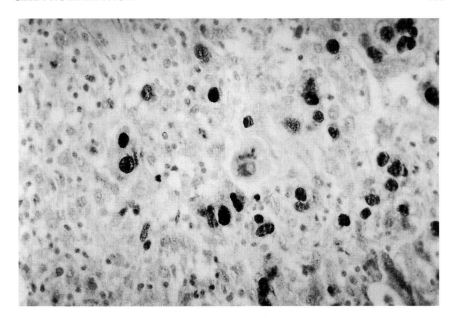

Fig. 6.9. Frozen section of a high-grade non-Hodgkin's lymphoma stained with the monoclonal antibody Ki-67, with labelling of the nuclei of proliferating cells. The proliferating nuclei are stained intensely but the morpohology is rather poor. Preparation courtesy of Miss Jane Oates, Histopathology Department, East Birmingham Hospital

sequences with respect to the Ki-67 antigen. It was also observed that all of these Ki-67 clones had 65–100% homology, which is remarkably high. When the Ki-67 cDNA was cloned into a bacterial expression vector, the antibody reacted with fusion proteins prepared only from bacteria where the insert was in the correct reading frame. Conversely, when it was present in the wrong reading frame, there was no binding. It can probably be safely assumed that the repetitious 62 bp unit encodes the Ki-67 epitope and, together with isolated cDNA clones, it has been used to characterize the Ki-67 gene. Both of these probes hybridize with mRNA (11.5 kbp or more) prepared from proliferating but not resting cells, and the polymerase chain reaction has subsequently been applied to the cloning and sequencing of a 6.3 kbp portion of the gene. Interestingly, this sequence appears to be unique and contains 16 repeats of the characteristic 62 bp sequence referred to above. The entire exon is of 6845 bp size.

The gene encoding Ki-67 appears to be located on the long arm of chromosome 10 (10q25), as has been shown by means of *in situ* hybridization using the 1095 bp sequence as a probe. Furthermore, the Ki-67 antigen has been localized at the ultrastructural level in the interphase of proliferating cells. The antigen is present in the outer parts of the nucleolus, especially in the granular component. When mitotic prophase commences, the antigen is seen in condensed chromatin and in metaphase on the chromatids. Later, it lies in the nucleoplasm. Thus, the localization differs from proliferating cell nuclear antigen and the major proteins associated with nucleolar organizer regions (see below), although numatrin (B23 protein) is also observed in the periribosomal zone.

Despite the acquisition of so much information regarding the Ki-67 antigen, the antibody generally available has suffered from a major drawback, namely the extreme lability of its antigen in the face of 'routine' processing to paraffin wax-embedded sections. However, there has recently been a major advance in the production of a novel antibody, designated MIB1, which was made by the immunization of mice with the central repeat area of Ki-67 antigen, expressed in *E. coli*. Prior to immunostaining, exposure of the sections to microwave irradiation is necessary (in citrate buffer) but morphological preservation is much better than that in frozen sections, and punctate areas of activity can be seen in proliferating nuclei. It is also possible to obtain highly satisfactory staining with the original Ki-67 antibody in this way (Fig. 6.10).

Proliferating cell nuclear antibody

Proliferating cell nuclear antibody (PCNA) was first detected by means of human autoantibodies from patients with systemic lupus erythematosus. These

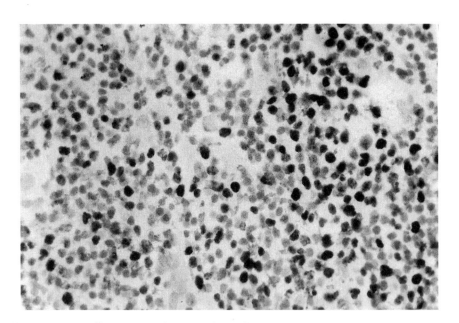

Fig. 6.10. Paraffin section of an anaplastic breast carcinoma, showing labelling of numerous nuclei with the antibody Ki-67 after prior microwave irradiation of the section. Note that the morphology is very well preserved. Preparation courtesy of Miss Jane Oates, Histopathology Department, East Birmingham Hospital

antibodies were found to react with ther nuclei of proliferating cells and we now have an increasing knowledge of the molecualr significance of PCNA. The antigen is of 36 kDa size and is an acidic non-histone protein. Cells cannot divide in its absence and it is an auxilliary protein for DNA polymerase δ. There are two populations of PCNA during the S phase of the cell cycle. One of these is nucleoplasmic in distribution, is present in low levels in resting cells and is readily extracted by detergents and organic solvents; the other is present in replication sites and is detergent-resistant. The latter co-localizes in space and time with incorporated BrdUrd and its level is lowered by the application of anti-sense oligonucleotides. It is this latter type of the protein which shows a large *proportional* rise of level during the cell cycle, even though there is only a two- to three-fold increase in *total* PCNA. The seminal importance of PCNA in DNA synthesis is underlined by the high level of evolutionary structural conservation of the protein. Thus, rat and human PCNA differ by only four amino acids, and even *Drosophila* PCNA has about 70% homology with the human protein.

Whether proliferating or not, the gene encoding for PCNA is transcribed; however, its mRNA accumulates only in proliferating cells. Growth factors, especially platelet-derived growth factor (PDGF), are of great importance in this context, since they stabilize PCNA mRNA and thus encourage translation, as the mRNA is otherwise unstable. This enables rapid response to stimulation and seems to be the result of splicing of intron 4. Oncogenes may also be involved in the control of PCNA mRNA levels. It is therefore apparent that control of PCNA production is under both transcriptional and post-transcriptional control.

Although autoantibodies were first used to detect PCNA, immunization of mice with *E. coli* bearing genetically engineered PCNA has led to the production of monoclonal antibodies to the protein. These include PC10, 19A2 and 19F4, which recognize different epitopes and demonstrate PCNA in routinely processed paraffin wax-embedded tissues (Hall *et al.*, 1990). It is of interest that these have rather different staining patterns to those seen with human autoantibodies and different sensitivities to histological processing. Variables such as tissue block size and type and duration are of importance, and sections must be produced and handled with great care. Two further problems may arise; firstly, PCNA may be seen in large amounts in normal or non-neoplastic cells adjacent to malignant tumours. This is the result of the formation of growth factors in the latter which induce the formation of PCNA in the former. This phenomenon has been shown experimentally in nude mice, when human cancer cells were injected into their kidneys or livers; there was a rise in the numbers of PCNA[+] cells in the adjacent hepatic and renal tissues without any increase in the numbers of S phase (i.e. genuinely proliferating) cells. The same effect can be shown by giving parenteral growth factors to rats. Secondly, difficulties may arise as a result of the relatively long half-life (> 20 h) of the protein. Thus, cells which have exited from the cell cycle may still

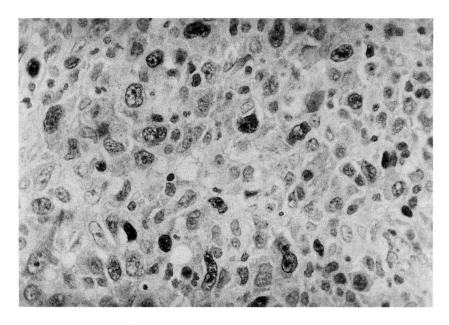

Fig. 6.11. A section of anaplastic carcinoma of the breast stained with the anti-PCNA antibody PC10. Most nuclei are reactive. Preparation courtesy of Miss Jane Oates, Histopathology Department, East Birmingham Hospital

express PCNA and the 'PCNA score' presumably overestimates the numbers of proliferating cells in tissue sections (Fig. 6.11).

As a result of the above observations, it is apparent that the interpretation of PCNA scores must be approached with considerable caution, since there are so many important variables to be considered. These include the method and duration of fixation, the clone of antibody used, the half-life of the antigen and the effects of growth factors. It is perhaps not in the least surprising that there are so many conflicting results with regard to staining with anti-PCNA antibodies and other measures of cell proliferation and in the context of studies of different tissues from different groups of researchers.

OTHER ANTIBODIES

Antibody to p105 antigen

Antibody to p105 in fact reacts with two proteins, with molecular weights of 105 and 41 kDa. The interpretation of this observation may be that there are monomeric and dimeric forms of the protein or that the smaller fragment is formed by partial proteolysis of the larger molecule. p105 is important in RNA synthesis and transport and in the regulation of the cell cycle. Teleologically, then, it is perhaps not surprising that it can be localized by means of immuno-electron microscopy to the interchromatin granules of the nucleus, the areas where RNA synthesis occurs. In the early phases of the cell cycle, p105 cannot be detected but there is a great rise in concentration during G_2 and mitosis itself. There is a concurrent change in the distribution of the protein during the cell cycle; thus, in interphase and early prophase p105 is seen in the interchromatin areas of the nucleus, but as the cycle progresses it is also present in the cytoplasm, after the nuclear membrane lyses. By means of immuno-cytochemistry it can readily be shown that p105 is present in proliferating but not in non-cycling cells. Furthermore, there is much greater expression in malignant cells, especially in those of high-grade malignancy. The antibody to p105 which is available commercially can be applied satisfactorily to 'routine' paraffin wax-embedded sections.

Antibody to p125 antigen

A protein, p125, so designated because of its 125 kDa size, has been shown in the nuclear matrix by the application of a monoclonal antibody. Increased levels of p125 can be seen in mitotic cells, being maximal at metaphase and anaphase, and its formation can be induced in lymphocytes by means of phytohaemagglutinin. In neoplasms and normal tissues, p125[+] cell numbers correspond approximately to those in S phase, although it may be that some non-proliferating cells may also express this protein. The antibody to p125 has yet to be fully evaluated in histopathology but can be applied successfully to paraffin sections.

Antibody to p145 antigen

A monoclonal antibody has been produced which reacts with a nucleolar protein in malignant cells but not with the cells of normal tissues. The protein is of 145 kDa size and the antibody was raised by immunizing mice with nucleolar extracts from HeLa cells. It would appear that the antibody has only been applied to cell preparations and to frozen tissue sections with immunofluorescent labelling.

Antibodies to DNA polymerase α and δ

DNA polymerase δ, which is an enzyme with fluctuating levels through the cell cycle, has peaks in the G_2 and M phases and tends to be associated with DNA polymerse α in its occurrence. It has 3' to 5' exonuclease activity and might be expected to reflect cell proliferation. DNA polymerase α is observed to peak, together with mRNA levels, at the transition from G_0 to G_1, just before the peak of DNA synthesis. Monoclonal antibodies to both enzymes, which are applicable only to frozen sections, have been described, and when antibodies to DNA polymerase α are applied to tumours proliferating cells are demonstrated. With antibodies to DNA polymerase δ, the picture is more complex, with staining of the nucleus, which spreads to the entire cell on mitosis. Unfortunately, these antibodies have yet to be fully investigated.

Antibody C_5F_{10}

A murine monoclonal antibody, C_5F_{10}, has been described and was produced by immunizing mice against 'tumour polysaccharide substance (TPS)-1–28', extracted from a human pulmonary carcinoma. The antibody demonstrates proliferating cells in paraffin wax-embedded sections and can be shown at the ultrastructural level to bind to multiple nuclear sites and sometimes to lysosomes. (The latter finding may explain the recent observation that the antibody may react with non-cycling cells.) It has been shown that the antibody does not react at the same sites as anti-PCNA or anti-tubulin. This is of importance, since tubulin (and vimentin) are known to be affected in their distribution by neoplastic transformation; similarly, so is the aggregation of tubules into microtubules with the occurrence of mitosis.

Antibody to protein p40

Antibody has been raised against a protein, Mr 40 000, designated p40, which is associated with nucleoli. It appears to be distinct from the major nucleolar organizer region-associated proteins and has been shown to be absent from normal human tissues but present in a range of malignant tissues, where it has a nucleolar distribution different from that of PCNA.

Antibody JC1

Recently a novel antibody, JC1, has been produced. It reacts with two antigenic components, of 123 and 212 kDa molecular weights, quite distinct from the proteins recognized by Ki-67. Furthermore, JC1 does not react with recombinant Ki-67 protein. The antibody is suitable for use with frozen sections only and has a similar but not identical staining pattern to that obtained with Ki-67.

Antibody KiS1

This novel antibody was raised against nuclear extracts from the promonocytic leukaemic cell line U937. It runs satisfactorily with paraffin sections and it has been related to prognosis in breast carcinoma. It reacts with a 160 kDa nuclear protein with a minor, 140 kDa element. Flow cytometric analysis shows that the kiS1 antigen increases in amount during the G_1 and S phases of the cell cycle, reaching levels four times greater than at G_2/M.

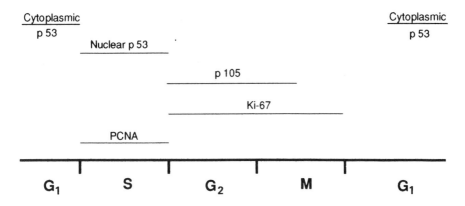

Fig. 6.12. Schematic diagram of the cell cycle with *peak level* expression of various cycle-related antigens

Antibody IND.64

This monoclonal antibody was produced by immunizing splenic cells from athymic nude mice grafted with a human lymphoblastic leukaemic cell line. The antibody reacts with proliferating cells in frozen sections and recognizes two antigenic components, with molecular weights of 345 and 395 kDa, assessed by immunoblotting. The antigens appear in late G_1 phase through to S,G_2 and M and is absent from G_0 and early G_1.

Fig. 6.12 summarizes the phases of the cell cycle at which some of the above molecules are expressed; clearly, there is considerable 'overlap' in the timing of their appearances. This may, in part, account for some of the apparent anomalies observed when these molecules are demonstrated in tissue sections.

ANTIBODIES TO NON-CYCLING CELLS

A prospect for the future lies in the possibility of generating antibodies to molecules characteristic of cells which are *not* proliferating. A likely candidate for such a marker is *statin*, a 57 kDa protein present in the nucleus. An antibody, S-30, has been raised against statin by immunizing mice with a detergent-resistant preparation from senescent, non-dividing human fibroblasts and this has localized statin to the nuclear envelope. The antibody demonstrates non-proliferating cells in frozen sections but the picture is complicated by the discovery of two different forms of the molecule. One is detergent-soluble and is seen in both dividing and resting cells; the other is detergent-insoluble and is observed only in non-dividing cells. The latter form of the antigen is lost progressively when cells are stimulated by serum to pass from G_0 to S phase of

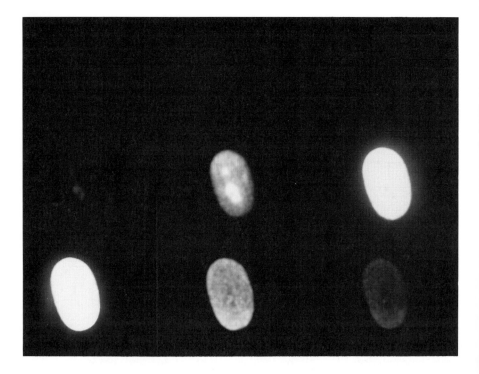

Fig. 6.13. A confocal scanning laser microscope preparation of a resting fibroblast, with six optical 'slices' showing the nuclear membrane distribution of labelling with the antibody BU31. Photograph courtesy of Drs R. Dover and C.P. Gillmore, Histopathology Unit, ICRF, Lincoln's Inn Fields, London

the cycle. It seems likely that other antibodies of this sort will become available and enable extended studies of cell populations in malignant and other tissues. An example antigen is a 30 kDa protein called *prohibitin*, which is present in non-cycling cells. Furthermore, screening of clones may reveal antibodies which are found empirically to bind to resting cells although their target antigen is not characterized. Such an antibody is BU31, which may possibly be directed against lamin and can be shown by means of scanning confocal microscopy to bind in the region of the nuclear membrane. It can be applied with success only to frozen sections, where it binds to resting cells (Figs 6.13 and 6.14).

IN SITU HYBRIDIZATION FOR HISTONE mRNA

Although not an immunocytochemical method as such, this novel approach has recently been used to demonstrate proliferating cells in paraffin sections (Fig. 6.15). Digoxigenin-labelled oligonucleotide probes have been used to

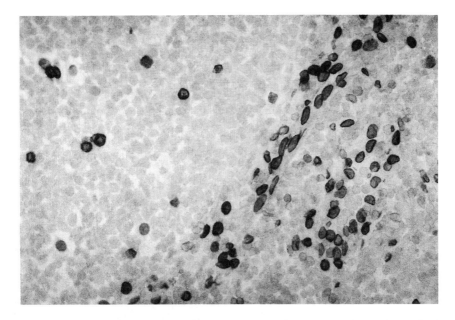

Fig. 6.14. A photomicrograph of a tonsillar lymphoid follicle, reacted with the antibody BU31; the positively stained cells are largely external to the follicle centre and correspond to cells which are known to be non-proliferating. Preparation courtesy of Miss Jane Oates, Histopathology Department, East Birmingham Hospital

demonstrate the histone mRNAs which are up-regulated in the S phase of the cell cycle. It has been shown that the method gives a significantly higher labelling index in high-grade than low-grade lymphomas, although the scores were lower than those obtained with Ki-67, probably reflecting the fact that these mRNAs are present for a shorter part of the cycle than the Ki-67 epitope.

CORRELATION BETWEEN IMMUNOLOGICAL MARKERS OF CELL PROLIFERATION AND CLINICAL STATUS

It is emphatically not the remit of a text such as this to describe the seemingly endless studies where immunohistochemistry for proliferation markers has been applied to clinical or histopathological diagnostic problems. For example, the reader should be able to appreciate the magnitude of the available data by referring to the information in Table 6.1, which gives as an example some of the conclusions reached in relation to numerous studies of tumour labelling by Ki-67 (Brown and Gatter, 1990). A list of similar length could now be given for studies with PCNA. Such investigations have sought to answer one or more of three main questions: (i) does a particular marker relate to grade or proliferative status of a specimen? (ii) does staining for the marker relate to survival/ prognosis in neoplastic disease? (iii) how can the data obtained from the

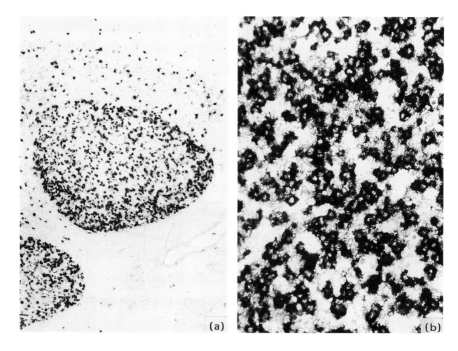

Fig. 6.15. *In situ* hybridization used to demonstrate histones in (a) a 'reactive' lymphoid follicle, where the the majority of basal cells are positive, and (b) a high-grade non-Hodgkin's lymphoma, with positive labelling of most cells. Preparations courtesy of Dr P.S. Colloby, Pathology Department, University of Leicester

application of one marker be related to those derived from the use of others? Certain generalizations can be made. It would be naive to expect the measurement of any one aspect of tumour extension to give an overall assessment of survival; indeed, cell proliferation is only one of many aspects of the malignant phenotype. Thus, enumerating the proliferating compartment of a particular cancer will often correlate well with the tumour grade but may or may not give an indication of patient prognosis. Other factors, such as cell motility, immune escape and secretion of proteolytic enzymes all have roles to play in the overall process of malignancy. Another limitation of the techniques generally used for the assessment of cell proliferation is that they give us a measure of proliferation *state*, not *rate*. The latter could potentially tell us much more about a specimen, since it would indicate the flux of cells entering and leaving what is usually a relatively static dividing pool. Furthermore, we do not yet have the ability either to detect or measure the numbers of *clonogenic cells* in histological material. These cells, as their name implies, are those that give rise to a proliferating clone, and an ability to demonstrate and quantify them in histological material would be highly desirable.

Table 6.1. Examples of the clinical application of Ki-67 in the assessment of malignant tumours

Tissue studied	Results and comments
Lymphoid	In general, high Ki-67 scores correlate with poor prognosis and high-grade malignancy in lymphomas. Good correlation with AgNOR scores
Breast	Ki-67 scores related well to recurrence after mastectomy, histological grade and mitotic activity. Most studies show no correlation with tumour size, lymph node involvement or oestrogen receptor status
CNS tumours	Histological grade in general reflected by Ki-67 counts although the latter did not always relate to survival data
Lung tumours	Ki-67 scores highest in oat cell carcinomas; lowest in carcinoids
Colorectal carcinomas	In general, poor correlation with histological type and other prognostic indices, including stage. High Ki-67 score in recurrent tumours
Hepatocellular carcinoma	Ki-67 score relates well with histological grade
Prostatic carcinoma	Ki-67 score related well to histological grade and clinical stage. With hormone therapy, the score decreases dramatically
Malignant melanoma	Tumour thickness correlates well with Ki-67 count
Uterine cervical carcinoma	No correlation between Ki-67 score and histological types and changes
Connective tissue tumours	Ki-67 index correlates well with histological grade and mitotic counts, as well as survival

The potential difficulties arising from the phenomenon of tumour cell heterogeneity have been outlined above and may, again, explain the lack of consistency of data from one study to another or between different specimens of the same tumour type. There may also be substantial variation in cell proliferation from areas which are, for example, at the growing edge of a neoplasm in relation to the deeper, relatively hypoxic areas. Tissue sampling, then, becomes a further consideration. The question of the relationship between data obtained from the use of different (immunocytochemical and non-immunocytochemical) techniques is complex. Certain 'markers', such as AgNOR scores (see Chapter 7) and Ki-67 counts generally correlate well, although conflicting results have recently been recorded between AgNOR numbers and PCNA counts. Likewise, there are papers describing high agreement between, say, S phase cells as assessed by DNA flow cytometry (Chapter 5) and AgNORs or Ki-67 labelling. When these data are further related to prognosis and survival, the picture becomes even more confused, doubtless for some or all of the reasons cited above. Nonetheless, the reader should not be discouraged from further studies of cell proliferation in tissues, since there is still much to be established and learned.

REFERENCES

Gatter KC, Brown NG, Trowbridge IS, Woolston RE and Mason DY (1983) Transferrin receptors in human tissues: their distribution and possible clinical relevance. *J. Clin. Pathol.*, **36**, 539–545.

Gerdes J, Dallenbach F, Lennert K, Lemke H and Stein H (1984) Growth fractions in malignant non-Hodgkin's lymphomas (NHL) as determined *in situ* with the monoclonal antibody Ki67. *Hematol. Oncol.*, **2**, 365–371.

Gratzner HG (1982) Monoclonal antibody to 5-bromo and 5-iodo-deoxyuridine: a new reagent for detection of DNA replication. *Science*, **218**, 474–475.

Hall PA, Levison DA, Woods Al *et al.* (1990) Proliferating cell nuclear antigen (PCNA) immunolocalisation in paraffin sections: an index of cell proliferation with evidence of de-regulated expression in some neoplasms. *J. Pathol.*, **162**, 285–294.

Lohka MJ, Hayes MK and Maller LJ (1988) Purification of maturation-promoting factor, an intracellular regulator of early mitotic events. *Proc. Natl Acad. Sci. USA*, **85**, 3009–3013.

Luca FC and Ruderman JV (1989) Control of programmed cyclin destruction in a cell-free system. *J. Cell Biol.*, **109**, 1895–1909.

McIntosh JR and Koonce MP (1989) Mitosis. *Science*, **246**, 622–628.

Pardee AB (1974) A restriction point for the control of normal animal cell replication. *Proc. Natl Acad. Sci. USA*, **71**, 1286–1290.

Pardee AB (1989) G1 events and regulation of cell proliferation. *Science*, **246**, 603–608.

FURTHER READING

Brown DC and Gatter KC (1990) Monoclonal antibody Ki-67: its use in histopathology. *Histopathology*, **17**, 489–503.

Ciba Foundation Symposium 170 (1992) *Regulation of the Eukaryotic Cell Cycle*. Chichester: Wiley.

Crocker J (1989) Review: proliferation indices in malignant lymphomas. *Clin. Exp. Immunol.*, **77**, 299–308.

Crocker J (1993) *Cell Proliferation in Lymphomas*. Oxford: Blackwell Scientific Publications.

Hall PA, Levison DA and Wright NA (1992) *Assessment of Cell Proliferation in Clinical Practice*. Berlin: Springer-Verlag.

Young S (1992) Dangerous dance of the dividing cell. *New Scientist*, **1824**, 23–27.

7 Interphase Nucleolar Organizer Regions

M. DERENZINI and D. PLOTON

It has been known for a long time that unusually large and irregularly shaped nucleoli characterize neoplastic cells. A series of light and electron microscopical studies have demonstrated that these nucleolar changes can be considered to be important cytopathological parameters for the diagnosis of malignancy (Busch and Smetana, 1970). However, changes of nucleolar morphology have not been of great importance in routine diagnostic tumour pathology as a consequence of the difficulty of quantifying them objectively.

In the past 15 years the structural–functional organization of the nucleolus in eukaryotic cells has been investigated extensively (Goessens, 1984; Derenzini *et al.*, 1990b). Data have been obtained which prompted histopathologists to re-evaluate the importance of the nucleolus in tumour pathology. A new parameter was defined that permits morphological changes of the nucleolus to be quantified objectively in routine cytohistological samples. This parameter concerns the distribution of the silver-stained interphase nucleolar organizer regions (NORs) (Crocker, 1990; Derenzini and Trerè, 1991a).

The NORs were first described as weakly staining chromatinic regions round which nucleoli reorganize during telophase. These regions correspond to the secondary constrictions of metaphase chromosomes which are located in the acrocentric chromosomes in man, i.e. chromosomes 13, 14, 15, 21 and 22. The NORs contain ribosomal genes, as revealed by *in situ* hybridization. A peculiar group of acidic proteins which have a high affinity for silver (AgNOR proteins) are also located in the NORs. These silver-stained proteins represent a good marker for ribosomal genes both in metaphase chromosomes and interphase nuclei (Figs. 7.1 and 7.2) (Howell, 1982).

During interphase the NORs are present in the fibrillar components of the nucleolus (Hernandez-Verdun, 1983). There is evidence that the distribution of interphase NORs in the nucleolar body is responsible for nucleolar morphology and that nucleolar changes can be effectively evaluated by considering the number and size of interphase NORs. In 1986 Ploton and co-workers succeeded in visualizing interphase NORs at the light microscope level in routine paraffin sections by applying a rapid, simple silver-staining method for

Molecular Biology in Histopathology. Edited by J. Crocker

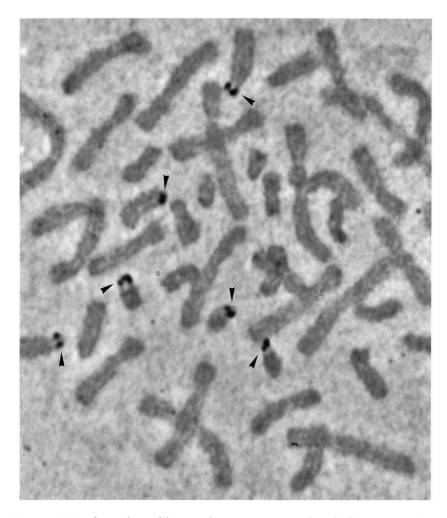

Fig. 7.1. Metaphase plate of human chromosomes stained with the one-step silver-staining method. No counterstaining. Phase contrast. Six acrocentric chromosomes show silver granules (arrowheads). Reproduced with permission from *Int. Rev. Exp. Pathol.*, **32**, 1991, Academic Press Inc.

the NOR proteins. By means of this procedure interphase NORs appear as well-defined black dots, distributed within the nucleolar body, which can be very easily quantified. Thereafter more than 400 papers have been published dealing with the importance of interphase AgNOR quantity in tumour pathology and for the diagnostic and prognostic characterization of cancers.

The aim of this chapter is to describe the structural–functional organization of interphase NORs in relation to the biology of cancer cells. The reason why interphase AgNOR number increases in cancer cells will be stressed for a better

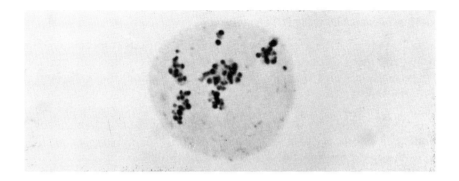

Fig. 7.2. Human tumour cell cultured *in vitro* stained with the one-step silver method. Numerous black dots are clustered in the nucleoli. × 2000

understanding of the importance that interphase AgNOR quantification may have for diagnostic and prognostic purposes in tumour pathology. The staining procedures for routine cytohistological samples will also be described, together with the methods for interphase AgNOR quantification. Since interphase NORs are nucleolar components, attention will first be focused on the architecture and molecular organization of the nucleolus.

THE NUCLEOLUS

MOLECULAR COMPONENTS

rRNA genes

In the nucleolus of one growing human cell about 400 ribosomal (r) RNA genes may be present. Although active rRNA genes may all be present within the nucleolus they only represent 1–2% of the nucleolar DNA disposed within the peri- and intra-nucleolar chromatin.

When isolated nucleoli are spread on a grid and subsequently observed in a transmission electron microscope, actively transcribed rRNA genes appear as the so-called 'Christmas tree' structures. These structures are randomly disposed and each is made up of a central axis (the rRNA gene itself) loaded with numerous RNA polymerase I molecules and surrounded by molecules of growing rRNA transcripts associated with ribonucleoprotein.

Each 'Christmas tree' corresponds to a single transcription unit (i.e. one transcribed rRNA gene containing the regions coding for the mature 18 S, 5.8 S and 28 S rRNA) and are separated from each other by a non-transcribed spacer (Hadjiolov, 1985).

Proteins associated with rRNA genes

It is estimated that the nucleolus contains at least 200 protein species which are histone and non-histone proteins, ribosomal proteins and pre-ribosomal particles and enzymes mainly controlling transcription and processing of rRNA transcripts.

The transcription of rRNA genes requires a set of proteins for the decondensation of DNA and for the transcription of the rRNA precursor. At present the more well-known proteins are RNA polymerase I, DNA topoisomerase I, protein factors associated with RNA polymerase I and nucleolin.

RNA polymerase I

RNA polymerase I is the specific polymerase for rRNA gene transcription. It is a heteromeric enzyme constituted by eight subunits with molecular weight (Mr) from 190 to 17.5 kDa. It is known that at least 2.5×10^4 molecules of RNA polymerase I are located within a diploid cell. Moreover, as observed on spread transcription units, 50 RNA polymerase I molecules are present on each micrometre of rRNA genes. Finally, each rRNA polymerase elongates rRNA transcript at a rate of around 40 nucleotides per second and requires only 10 min to synthesize one pre rRNA molecule.

Antibodies raised against RNA polymerase I precipitate a protein complex with about 13 proteins with Mr ranging from 210 to 12.5 kDa. Several co-factors of RNA polymerase I are found among these proteins. A high-Mr RNA polymerase I subunit has been suggested to be an AgNOR protein.

Topoisomerase I

This protein may also be considered an elongation factor for RNA polymerase I because it suppresses the stress in the DNA helix caused by progression of RNA polymerase I along the rRNA gene. Immunocytochemistry demonstrated that topoisomerase I is associated with RNA polymerase I and it has also been shown that topoisomerase I molecules are present on transcribed regions of rRNA genes and absent on non-transcribed spacers.

Nucleolin or C23 protein

Nucleolin is the major nucleolar phosphorylated protein of exponentially growing cells. This protein (100 kDa) contains three domains, among which two play important functional roles in the transcription of rRNA molecules. With its acidic N-terminal domain, nucleolin binds to histone H1 and induces the decondensation of rRNA genes, facilitating their transcription by RNA polymerase I. With its central domain, nucleolin binds to rRNA transcripts during their transcription and the first steps of their processing. It was also demonstrated that nucleolin is a preferential substrate of a nucleolar casein kinase NII which phosphorylates serine residues located near the N-terminal sequence (topoisomerase I and a subunit of RNA polymerase I are also phosphorylated by casein kinase NII). Nucleolin is the only nucleolar protein which has been clearly identified as an AgNOR protein (Roussel *et al.*, 1992).

Proteins associated with rRNA processing

Numatrin or B23 protein

This is a hexameric 38 kDa phosphoprotein which shows a high binding capacity with rRNA. It is associated with pre-ribosomal particles during the late steps of their organization. During serum deprivation or during inhibition of rRNA transcription by actinomycin D, the level of B23 protein correlates with cell growth: for example, numatrin synthesis and quantity is elevated in growth factor-stimulated 3T3 fibroblasts and in lectin stimulated T lymphocytes. These findings suggest that numatrin is associated with receptor-mediated induction of mitosis in various normal and cancerous cell types. There is some evidence indicating that B23 protein may be an AgNOR protein.

Fibrillarin

This is a basic 34 kDa protein rich in N^G, N^G-dimethyl arginine and glycine clustered at the N-terminus of the molecule.

This molecule is of particular interest because it is associated with U3 RNA within a ribonucleoprotein particle. U3 small nucleolar RNA is associated with pre-RNA processing (cleavage of the 5.8 S rRNA is mediated by U3 RNA). Recently it has been shown that fibrillarin and U3 RNA are associated with U8 and U13 small nucleolar RNA, which also play a role in processing pre-RNA molecules.

Other nucleolar proteins

Several proteins have been isolated and characterized from nucleoli of cancerous cells. They are specifically found within nucleoli of malignant cells and are absent within normal cells. These are p145, p125, p120 and p40 proteins. Another protein, also associated with proliferation, is PCNA (proliferating cell nuclear antigen). It is a co-factor of DNA polymerase, which is localized within both the nucleolus and the nucleoplasm (see Chapter 6).

Autoantibodies reactive with nucleolar components are very frequently found in sera from patients with scleroderma. These autoantibodies were used first to acquire more data regarding the organization of the nucleolus and second to identify new proteins involved in ribosome biogenesis. Antigens

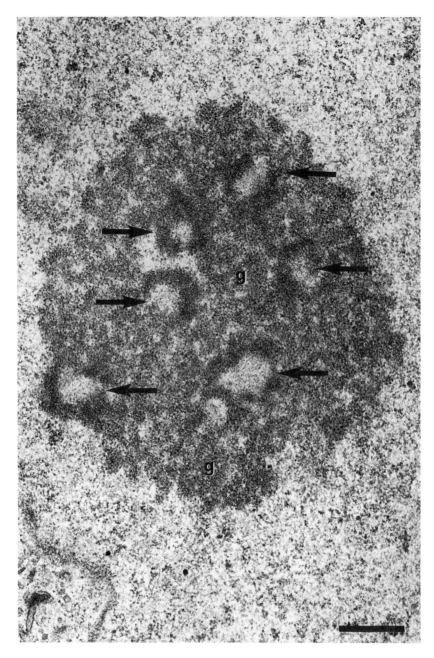

Fig. 7.3. Nucleolus of a TG cell. Six fibrillar centres are present, with the closely associated dense fibrillar component (arrows). g, granular component. Aldehyde and osmium fixation; uranium and lead staining. × 35 000. Bar 0.5 μm. Reproduced from Derenzini and Trerè (1991a) by permission of Springer–Verlag

recognized by these antibodies correspond to proteins with M_r ranging from 116 to 20×10^3. Some of these antibodies recognize antigens whose chemical nature is known, such as RNA polymerase I and fibrillarin. However, others such as polymyositis–scleroderma antigen immunoprecipitates several poly-peptides with an M_r from 110 to 20 kDa whose function is still unknown.

STRUCTURE AND FUNCTION

The nucleolus is a well-defined structural–functional unit of the interphase cell nucleus, in which ribosomal genes are located and ribosomal biogenesis occurs. At the electron microscope (EM) level, in routine thin sections stained with uranium and lead, the nucleolus is composed of three main components: the fibrillar centres, the dense fibrillar component and the granular component. The fibrillar centres are roundish structures, highly variable in size, characterized by light electron-opaque fibrillar material. The dense fibrillar component, com-posed of highly clustered very thin fibrils, is more electron-opaque than the fibrillar centres. The dense fibrillar component is usually found surrounding and intimately associated with the fibrillar centres; sometimes, however, it may be located at a distance from the fibrillar centres without any apparent spatial relationship with them. The granular component is always found more externally with respect to the fibrillar components (Fig. 7.3). The distribution of these nucleolar components changes from cell to cell and in the same type of cell from the resting to the proliferating state, thus producing a high variability of nucleolar size and shape.

Cytochemical, immunocytochemical and *in situ* hybridization ultrastructural techniques, together with high-resolution autoradiography, have enabled the molecular composition and the structural–functional relationship of the nucleolar components to be defined. Application of the silver staining method for NOR proteins at the EM level has shown that the only cell structures stained are the fibrillar centres and the closely associated dense fibrillar components (Fig. 7.4). A Feulgen-like procedure, using an osmium–ammine complex as an electron-opaque marker, demonstrated that in the fibrillar centres and their immediately associated dense fibrillar component DNA structures were also present. This DNA was composed of filaments character-ized by a completely extended, 'open' configuration. *In situ* hybridization techniques indicated that this DNA was rDNA.

These observations led to the conclusion that the nucleolar fibrillar components were the interphase counterpart of metaphase NORs. They were the only nucleolar structures where ribosomal genes and AgNOR proteins were co-located. Indeed, in interphase cells ribosomal genes were not exclusively located in the fibrillar components of the nucleolus, but were also found in the extranucleolar compartment, where they have a highly compact structure. Immunocytochemical studies have also shown that the molecules necessary for ribosomal transcription, i.e. RNA polymerase I and

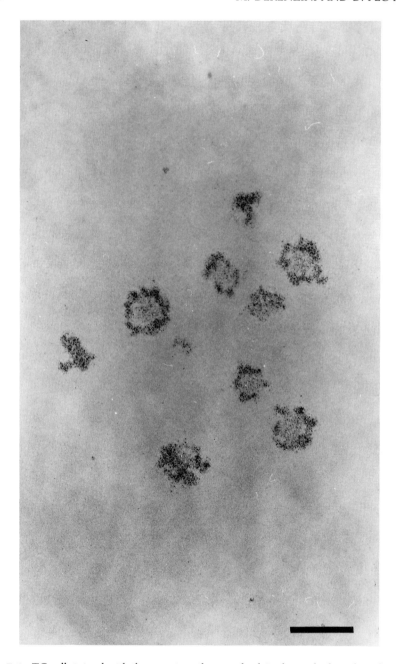

Fig. 7.4. TG cell stained with the one-step silver method. In the nucleolus, silver deposits are located exclusively in the fibrillar centres and, to a larger extent, in the associated dense fibrillar component. No counterstaining. × 35 000. Bar 0.5 μm. Reproduced from Derenzini and Trerè (1991a) by permission of Springer–Verlag

Table 7.1.

Nucleolar components	Molecules identified	Main function of the component
Fibrillar centres	rDNA AgNOR proteins RNA polymerase I	Stock for proteins
Dense fibrillar component closely associated with fibrillar centre	rDNA AgNOR proteins RNA polymerase I Topoisomerase I rRNA (45 S) [³H]uridine incorporation Nucleolin Fibrillarin	Transcription and initial processing of rRNA
Granular component	B 23 Nucleolin rRNA 18 and 28 S	Processing of rRNA

topoisomerase I, were located in the fibrillar nucleolar components. Autoradiographic studies and correlated biochemical analysis indicated that the newly transcribed 45 S pre-ribosomal RNA molecules were accumulated in the dense fibrillar component and that the granular component contained rRNA molecules resulting from 45 S RNA processing. In Table 7.1 the main molecules so far identified, their localization in the nucleolus and the major functions of the components are summarized. As far as the dense fibrillar component is concerned, only the molecular composition and related function of that portion, which is closely associated with the fibrillar centre, is reported. There is increasing evidence that the molecular composition and function of that portion of dense fibrillar component which is located at a distance from the fibrillar centres is quite different from that surrounding the fibrillar centres. For example, the former does not contain either rDNA or AgNOR proteins.

Assigning functions to the nucleolar components, it can be concluded that the interphase NORs represent the 'heart' of the nucleolus. In these areas ribosomal genes are located together with all proteins necessary for transcription such as RNA polymerase I, topoisomerase I and AgNOR proteins. In the confines of the interphase NORs, rRNA transcription takes place.

INTERPHASE NORs AND NUCLEOLAR STRUCTURAL ORGANIZATION

The size and shape of the nucleolus is highly variable, depending on cell activity. In resting cells such as human circulating lymphocytes, the nucleolus is

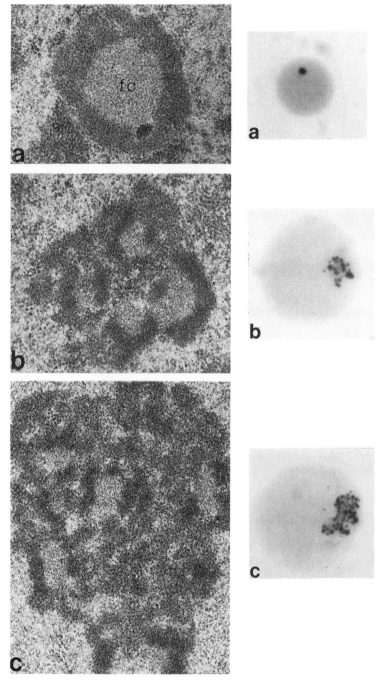

Fig. 7.5   Fig. 7.6

very small with a very simple structural organization. At the EM level the nucleolus of a resting lymphocyte is composed of a solitary, large fibrillar centre surrounded by a rim of ribonucleoprotein structures (ring-shaped nucleolus). If resting lymphocytes are stimulated to proliferate by phytohaemagglutinin, the nucleolus progressively enlarges and the distribution of nucleolar components becomes more and more intricate (Fig. 7.5). The number of fibrillar centres greatly increases but, at the same time, the size is greatly reduced. Each fibrillar centre is surrounded by a rim of dense fibrillar component frequently intermingled with granules. These rims, as a consequence of the proximity of the fibrillar centres, come in contact with each other, giving rise to continuous cord-like structures (nucleolonema organization).

Changes of nucleolar morphology are very easily identifiable at the EM level, whereas they are detected with great difficulty with the light microscope, especially when routine histological sections are considered. Small nucleoli are frequently undetected (leading to the current, wrong conviction that some cells lack nucleoli!) and larger nucleoli, independent of their size and shape, are generally defined as 'prominent nucleoli'. On the other hand, a series of ultrastructural studies have demonstrated that different nucleolar morphological patterns can be distinguished from each other on the basis of distributional variations of the fibrillar centres. The total volume occupied by the fibrillar centres in nucleoli was closely correlated with the volume of the whole nucleolus. This relationship can be explained by considering that the fibrillar centre contains the rDNA sequences coding for rRNA and represents the core round which the nucleolus with the ribonucleoprotein components is built up. Changes in the size and number of fibrillar centres directly reflect the quantity and distribution of the ribonucleoprotein components and, ultimately, the nucleolar volume and shape. Measurement of size and number of fibrillar centres has therefore been used to estimate nucleolar morphological differences, in an objective and reliable manner, in both normal and cancer cells at the ultrastructural level.

The importance of quantifying nucleolar morphological changes using a parameter relative to fibrillar centre distribution was greatly increased after Ploton and co-workers (1986) applied the one-step silver method for staining AgNOR proteins at the light microscope level and succeeded in visualizing the

Fig. 7.5 (*opposite page, left*). Ultrastructural aspects of control (a), and phytohaemagglutinin-stimulated human lymphocytes (b and c). Note the different nucleolar morphology as a consequence of the different distributions of the interphase NORs (fibrillar centres (fc) plus dense fibrillar component). × 37 000. Reproduced with permission from *Int. Rev. Exp. Pathol.*, **32**, 1991, Academic Press Inc.

Fig. 7.6 (*opposite page, right*). Smeared preparations of the lymphocyte samples shown in Fig. 7.5. One-step silver method. × 1200. Reproduced with permission from *Int. Rev. Exp. Pathol.*, **32**, 1991, Academic Press Inc.

fibrillar centres in routine cytological and histological samples. Indeed, as previously stated, the AgNOR staining method, apart from pinpointing the fibrillar centres, also stains the dense fibrillar component surrounding the fibrillar centre itself. The complex fibrillar centre plus the intimately associated dense fibrillar component, stained by silver, constitute the interphase AgNOR. In preparations stained according to the method of Ploton and co-workers (1986), the interphase NORs appear, at the light microscope level, as well defined black dots. As shown in Fig. 7.6 changes in the nucleolar shape and size can be evaluated easily by considering the number and the spatial distribution of the silver-stained dots. Indeed, it has been demonstrated that the total area occupied by the interphase AgNORs in the nucleolus is related to the nucleolar area.

Interphase AgNOR distribution can therefore be considered to represent a reliable parameter for evaluating changes in nucleolar morphology and activity at the light microscope level in routine cytological and histological samples.

SILVER STAINING PROCEDURE FOR NORs

NORs are characterized by their high affinity for silver (Howell, 1982). After silver staining, NORs appear as black dots of metallic silver (AgNOR dots) around 0.5 to 1 μm in diameter, localized within secondary constrictions of metaphase chromosomes or within nucleoli (Figs 7.1 and 7.2).

Originally, silver-staining procedures were performed in two or three steps. At each step the possibility of increasing the non-specific background staining limited these methods which were applied rarely and with difficulty to routine histology.

In 1980, Howell and Black proposed a very reproducible one-step method for staining NORs within metaphase preparations. By modifying this procedure, Ploton and co-workers (1986) obtained a reproducible technique for AgNOR staining of cytohistological samples for routine investigation both at the light and EM levels.

Ultrastructural studies demonstrated that after this staining very small silver deposits were localized specifically within the fibrillar centres and dense fibrillar components. Moreover, the absence of background staining either within the nucleoplasm or the cytoplasm speaks in favour of the specificity of this procedure.

MOLECULAR COMPONENTS RESPONSIBLE FOR SILVER STAINING

Stainable silver material resists RNAse, hydrochloric or trichloroacetic acid treatment but not trypsin digestion. This indicates that the molecules stained are proteins. Several proteins are candidates for this argyrophilia: nucleolin, numatrin, a subunit of RNA polymerase I and another still unspecified

'AgNOR protein'. Until recently, only nucleolin has been identified as an AgNOR protein (Roussel *et al.*, 1992). Protein-bound sulphydryl and disulphide groups are probably responsible for silver staining. The argyrophilic nature of nucleolin is related to its N-terminal domain, i.e. the domain which binds it to rDNA, and is particularly enriched with acidic amino acids. No correlation has been found between the level of phosphorylation of nucleolin and its silver stainability. Although nucleolin is found both in the dense fibrillar and in the granular component, the absence of silver staining in the latter may be explained by proteolytic cleavage of nucleolin which leads to the loss of its N-terminal domain.

THE ONE-STEP SILVER-STAINING METHOD

In the original technique of Howell and Black (1980) staining is performed at 70 °C for only 2 min. In order to obtain more reproducible results without non-specific background the following are recommended:

(1) Use a lower temperature.
(2) Use a longer staining time.
(3) Rinse with thiosulphate solution in order to wash out non-specific deposits and to prevent further sensitivity of the slide to light.

Two solutions are needed: (A) 2% solution of gelatin dissolved in ultra-pure water to which formic acid is added to make a 1% solution; (B) 50% silver nitrate solution in ultra-pure water. Staining solution is obtained instantly by rapidly mixing one part of solution A with two parts of solution B into a glass cylinder. It is better to avoid direct sunlight exposure during the staining procedure. After staining, the solution is poured off and slides washed in several baths of ultra-pure water, placed in a 5% thiosulphate solution for 10 min and washed again in several baths of ultra-pure water. In Table 7.2 the staining procedures used for differently processed samples are reported.

TECHNIQUES FOR AgNOR QUANTIFICATION

The first method applied by pathologists to quantify interphase AgNORs in routine histopathological samples was the 'counting method' (Crocker *et al.*, 1989). This consists of the counting of the number of single silver-staining dots per cell, evaluating the sample directly at the microscope and carefully focusing through the section thickness at high magnification (1000 ×). Although this 'counting method' has shown that interphase AgNOR quantification can be very helpful in tumour pathology, it presents some substantial handicaps from a technical point of view: it is time-consuming, tedious and, moreover, rather subjective and may have low reproducibility. When, in particular, interphase AgNORs are clustered together or partially overlapping, as often happens in

Table 7.2.

Smears and chromosomes
1. Cytological samples are air-dried or fixed in Merckofix or 95% ethanol
2. Post-fix for 10 min in Carnoy's solution (absolute ethanol–glacial acetic acid 3:1 v/v) and then rehydrate
3. Stain with silver solution in the dark at a constant temperature of 37 °C for 13–15 min
4. Dehydrate and mount as routine

Frozen histological samples
1. After drying, fix the sections in 95% ethanol or 'methcarn' (methanol–chloroform–glacial acetic acid 6:3:1 v/v/v)
2. Post-fix for 10 min in Carnoy's solution and then rehydrate
3. Stain with silver solution in the dark at a constant temperature of 37 °C for 13–15 min
4. Dehydrate and mount as routine

Histological samples fixed in 95% ethanol or in other alchol-based fixatives, and routinely paraffin-embedded
1. Dewax the sections in xylene
2. Post-fix for 10 min in Carnoy's solution and then rehydrate
3. Stain with silver solution in the dark at a constant temperature of 37 °C for 12–14 min
4. Dehydrate and mount as routine

Histological samples fixed in 10% buffered formalin, and routinely paraffin-embedded
1. Dewax the sections in xylene
2. Rehydrate the sections, avoiding any post-fixation
3. Stain with silver solution in the dark at a constant temperature of 37 °C for 16–20 min
4. Dehydrate and mount as routine

Tissues or cell pellets for ultrastructural studies
1. Samples are fixed in glutaraldehyde (2% in phosphate-buffered saline) for 10 min
2. Post-fix for 10 min in Carnoy's solution
3. Stain samples (in small Petri dishes) with silver solution in the dark at a constant temperature of 37 °C for 12–16 min
4. Dehydrate in graded ethanol and embed in Epon or LW White

---

[a]The given choice for temperature and time depends on the facilities of the laboratory. If there are fluctuations of room temperature due to external conditions (seasons, country, etc.) it is better to choose a higher temperature which may be easily obtained and controlled within an oven (e.g. 37 °C). When this temperature is fixed, the appropriate time for staining which will give optimal results without background has to be found. When temperature and time length are chosen, then reproducible results are obtained for given samples.
[b]When good AgNOR staining has been performed, several well-separated black granules are localized within the nucleoli. Some granules may appear within the nucleoplasm: they probably correspond to small nucleoli or to coiled bodies. Clumps of chromatin should be unstained. If a brownish colour of chromatin appears, it means that the specimen is over-stained and that a shorter time for staining must be used. It must also be noted that sometimes connective tissue is silver-stained and that brownish deposits appear on collagen bundles. The reason for this is unclear. However, a short hydrolysis with 5 N hydrochloric acid for 10 min before staining should eliminate this unspecific staining.
[c]The silver-stained samples may be counterstained with methyl green for a clear identification of nuclei at transmitted light microscopy. A specific fluorochrome such as chromomycin A3 may be helpful for identifying nuclei and mitotic chromosomes with fluorescence microscopy.

malignant cells, the discrimination of each single dot is highly dependent on the observer's experience and counting criteria. It must also be said that simply counting AgNOR numbers does not take into consideration the different sizes of the silver-stained dots.

In order to overcome these drawbacks, several groups (Rüschoff *et al.*, 1990; Derenzini and Trerè, 1991a; Ploton *et al.*, 1992), working simultaneously, developed specific computer-assisted image analysis systems for the automatic or semi-automatic evaluation of the areas occupied by the silver-stained structures. Compared with the counting method, morphometric analysis presents several advantages: it is easy and rapid to execute, objective and highly reproducible.

Irrespective of the method used for interphase AgNOR evaluation, three factors greatly influence NOR stainability and, consequently, quantification: these are the fixative employed, and the temperature and time of the staining reaction. The use of different fixatives greatly affects the NOR stainability, as demonstrated by numerous studies. Samples fixed with formalin-containing solutions (buffered formalin or Bouin's fluid) show smaller AgNORs than samples fixed with alcohol-containing solutions (absolute ethanol or 'meth-carn' solution). The temperature and time of the staining reaction are inversely related to each other: the higher the temperature, the shorter the time required for NOR silver staining. When the staining reaction is prolonged beyond the time for selective visualization of NORs, all the other nucleolar structures are progressively stained, until the whole nucleolus appears homogeneously stained by silver. In order to obtain comparable data between different laboratories it is essential that the same fixative and the same staining protocol are used. However, if differently fixed or stained samples must be compared, quantitative evaluation can be performed by image analysis using the tumour-associated lymphocytes or stromal cells as an internal control. Different studies (Ruschoff *et al.*, 1990; Derenzini and Trerè, 1991b) have demonstrated that while measurements of interphase AgNOR areas were highly variable for the same tumour samples depending on the fixative employed and the temperature and time of the staining reaction, the ratio between the mean interphase AgNOR area of cancer cells and lymphocytes or stromal cells (NOR index) was found to be the same. The use of lymphocytes or stromal cells as an internal control for the standardization of AgNOR evaluation in cancer tissues is made possible by the fact that their mean AgNOR area is almost constant in human tumours, independent of sex and age.

INTERPHASE AgNORs IN TUMOUR PATHOLOGY

The possibility of an effective quantification of nucleolar changes by evaluating the interphase AgNOR distribution at the light microscope level has, in the past few years, greatly revalued the importance of nucleolar morphology in tumour pathology. In order to have a better understanding of the advantages and limitations of interphase AgNOR quantification as a diagnostic and prognostic parameter in tumour pathology we will first report (i) what is already known about the relationship between interphase AgNOR quantity and those structural–functional changes in cancer cells which can determine modification of the interphase AgNOR distribution, and (ii) the relationship between interphase AgNOR quantity and cell proliferation rate.

STRUCTURAL–FUNCTIONAL CHANGES WHICH MAY INFLUENCE INTERPHASE AgNOR DISTRIBUTION IN CANCER CELLS

As previously stated, interphase AgNORs contain active ribosomal genes and newly transcribed ribosomal ribonucleoprotein particles. Two different structural–functional characteristics of transformed cells can reasonably be considered important in determining the quantitative changes of interphase AgNORs in cancer cells. They are an increased number of acrocentric chromosomes carrying the NORs and an increased nucleolar metabolic activity for producing a larger quantity of ribosomes.

Ploidy

The relationship between the number of interphase and metaphase NORs has been clearly defined. Data obtained using various cell types have indicated the absence of a numerical relationship between interphase and metaphase NORs. In activated cells the number of interphase AgNORs greatly exceeds that of the AgNORs present in metaphase chromosomes, whereas in resting cells (like the human circulating lymphocytes) the interphase AgNOR number ranges from one to two and is markedly lower than that of silver-stained acrocentric chromosomes. On the other hand, even if the number of interphase AgNORs is not strictly related to the number of metaphase AgNORs, the possibility may exist that an increased number of metaphase AgNORs could result in a greater number of interphase AgNORs. Since hyperdiploidy is the most frequent change in chromosome number that distinguishes tumour from non-tumour cells, this point is of particular interest in defining those mechanisms involved in causing an increased number of interphase AgNORs in cancer cells. Studies carried out on established tumour cell lines of different origin have shown a positive correlation between the number of acrocentric chromosomes with AgNORs and the whole chromosome number. These results might lead to the supposition that the high number of interphase AgNORs present in cancer cells might be the consequence of an increased number of acrocentric chromosomes carrying the silver-stained NORs. However, a quantitative study on the relationship between the number of interphase and metaphase AgNORs in neuroblastoma cell lines demonstrated that the number of interphase AgNORs was not related to the number of metaphase AgNORs. It can therefore be concluded that ploidy does not directly influence the expression of interphase AgNORs. On the other hand, conflicting results have been reported about the relation between DNA content and interphase AgNOR quantity in human tumour samples. Even if there is increasing evidence indicating that the two parameters are not directly related, tumours with high DNA content are frequently associated with high interphase AgNOR values. This fact should be considered in the light of a greater metabolic activity of hyperdiploid versus euploid cells that could determine an increased interphase AgNOR expression.

rRNA transcriptional activity

There is general agreement on the fact that the number of interphase AgNORs is related to rRNA transcriptional activity. In late telophase, when rRNA synthesis starts again after the mitotic block, only one NOR is present in the nucleolus. The progressive increase in ribosome biogenesis which occurs during G_1 and S phases is associated with a gradual increase in the number of interphase NORs. The same occurs in resting lymphocytes stimulated by phytohaemagglutinin. On the contrary, the number of interphase NORs progressively decreases during differentiation and reduction of transcriptional activity, such as in epithelial cells of the intestine during cell migration from the germinative zone of the crypt to the top of the villus. Reduction of interphase NOR number has been constantly detected after drug-induced inhibition of rRNA synthesis.

The strict relationship observed between interphase AgNOR number and rRNA transcriptional activity is consistent with the finding that transcription-ally active ribosome genes are present only with interphase AgNORs, and that rRNA synthesis takes place within their confines. Can we therefore conclude that the quantity of interphase AgNORs merely reflects the entity of rRNA transcriptional activity? Data indicate that this is not the case. Cultured cells with the same level of rRNA synthesis, but characterized by different doubling times, had different amounts of interphase AgNORs. Stimulation of rRNA synthesis in rat hepatocytes by cortisol injection did not induce an increase of interphase AgNORs to the same extent as that found in regenerating hepatocytes with similar rRNA transcriptional activity. Thus, even if stimulation of ribosomal biogenesis is associated with an increased number of interphase AgNORs this is not the only event which determines changes of interphase AgNOR quantity. There is, indeed, some evidence indicating that the increased amount of interphase AgNORs in proliferating cells might also be related to structural changes of ribosomal chromatin occurring before ribosomal gene duplication. As previously stressed, in interphase NORs the AgNOR proteins are always associated with ribosomal chromatin with an extended configuration. In cells stimulated to proliferate the newly synthesized

AgNOR proteins become progressively associated with compact ribosomal genes and by decondensing them give rise to new interphase AgNORs. Decondensation of compact ribosomal genes is probably necessary, apart from for the increase rRNA transcription request, also for ribosomal genes duplication.

INTERPHASE AgNOR QUANTITY AS A PARAMETER OF CELL KINETICS

In cells stimulated to proliferate the quantity of interphase AgNORs progressively increases from early G_1 phase, reaches a maximum value during S phase and remains constant up to the late G_2 phase. Consequently, it is not surprising to observe that proliferating tissues exhibit interphase AgNOR values greater than non-proliferating ones. A series of data have been produced which show that interphase AgNOR quantity is related to the proliferation indexes determined in tissue samples by well-established parameters of cell kinetics (Table 7.3). Of particular interest is the finding that the interphase AgNOR amount is directly related to the duplicating activity, irrespective of the type of tissue (Figs 7.7 and 7.8). Tumours of different origin exhibit a highly significant correlation between the interphase AgNOR quantity and cell proliferation rate evaluated by Ki-67 immunostaining and/or bromodeoxyuridine (BrdUrd) *in vitro* labelling. It appears therefore that quantification of interphase AgNORs can actually represent a useful tool for cell kinetics evaluation. The importance of the quantity of interphase AgNORs as a parameter of cell proliferation rate has been greatly increased after the observation that the amount of interphase AgNOR has been shown to be strictly related to the *rapidity of cell proliferation* (Derenzini et al., 1990a). This was clearly demonstrated using tumour cell lines of different origin cultured *in vitro*, characterized by different doubling times. It was shown that cell lines with a difference of only 4 h regarding their doubling time had a significant difference in interphase AgNOR quantity. How can we explain the strict relationship between interphase AgNOR quantity and cell doubling time? Interphase AgNOR accumulation in the cell entering the mitotic cycle is very likely related both to an increased request of ribosomal biogenesis and decondensation of compact ribosomal genes for the following duplication. Rapidly proliferating cells have to accomplish these functions in a shorter time than slowly proliferating cells. Consequently, the number of interphase AgNORs must be greater in rapidly than in slowly dividing cells and the quantity of interphase AgNORs strictly related to the rapidity of cell proliferation.

Among the methods proposed for quantifying the cell proliferation rate, the AgNOR parameter arrived last. However, its evaluation appears to be highly recommended for the following reasons with respect to the other methods for measuring cell kinetics:

Table 7.3. Correlation between interphase AgNOR quantity and cell proliferation

Tumour	Reference	Correlation
AgNOR and Ki-67		
Non-Hodgkin's lymphomas	Hall *et al.* (1988) *Histopathology*, **12**, 373–381	Significant
	Okabe *et al.* (1991) *Anticancer Res.*, **11**, 2031–2035	Significant
Breast carcinomas	Dervan *et al.* (1989) *Am. J. Clin. Pathol.*, **92**, 401–407	Significant
	Raymond *et al.* (1989) *Hum. Pathol.*, **20**, 741–746	Significant
	Rüschoff *et al.* (1990) *J. Cancer Res. Clin. Oncol.*, **116**, 480–485	Significant
	Di Stefano *et al.* (1991) *Cancer*, **67**, 463–471	Significant
	Sasaki *et al.* (1992) *Oncology*, **49**, 147–153	Significant
Gliomas	Hara *et al.* (1990) *Surg. Neurol.*, **33**, 320–324	Significant
Brain tumours	Plate *et al.* (1990) *Acta Neurochir. Wien*, **104**, 103–109	Significant
Lung carcinomas	Soomro and Whimster (1990) *J. Pathol.*, **162**, 217–222	Insignificant
Gastric carcinomas	Kakeji *et al.* (1991) *Cancer Res.*, **51**, 3503–3506	Significant
Soft tissue sarcomas	Kuratsu *et al.* (1991) *Int. J. Cancer*, **48**, 211–214	Significant
Group of tumours of different origin	Trerè *et al.* (1991) *J. Pathol.*, **165**, 53–59	Significant
AgNOR and percentage of S-phase cells by BrdU incorporation		
Meningiomas	Orita *et al.* (1990) *Neurosurgery*, **26**, 43–46	Significant
Hepatocellular rat carcinomas	Tanaka *et al.* (1989) *Jpn. J. Cancer Res.*, **11**, 1947–1951	Significant
Group of tumours of different origin	Trerè *et al.* (1991) *J. Pathol.*, **165**, 53–59	Significant
Breast carcinomas	Sasaki *et al.* (1992) *Oncology*, **49**, 147–153	Significant
Acute leukemia	Nakamura *et al.* (1992) *Acta Haematol.*, **87**, 6–10	Insigificant
AgNOR and percentage of S-phase cells by DNA flow cytometry		
Non-Hodgkin's lymphomas	Crocker *et al.* (1988) *J. Pathol.*, **154**, 151–156	Significant
Breast carcinomas	Giri *et al.* (1989) *J. Pathol.*, **157**, 307–313	Postive trend
	Mourad *et al.* (1992) *Cancer*, **69**, 1739–1744	Significant
Gastric carcinomas	Rosa *et al.* (1990) *Histopathology*, **16**, 614–616	Positive trend

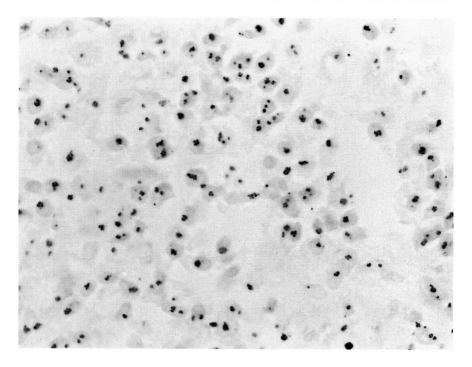

Fig. 7.7. Low proliferating carcinoma of the breast (BrdUrd labelling index, 2.6) stained with the one-step silver method. Carnoy fixation. × 360. Reproduced from Derenzini and Trerè (1991a) by permission of Springer–Verlag

(1) Visualization of interphase AgNORs is very rapidly obtained (10–20 min) using paraffin-embedded, routinely processed tissue sections. Other methods require fresh tissue (flow cytometry) or frozen sections (Ki-67 immunostaining) and are much more time-consuming.

(2) The interphase AgNOR quantity is the only parameter that indicates the *rapidity of cell proliferation* in routinely processed samples. Ki-67 and proliferating cell nuclear antigen (PCNA) methods reveal cells which entered the mitotic cycle. Visualization, on tissue sections, of cells which have previously incorporated BrdUrd into nuclear DNA, allows the number of cells in S phase to be determined, but not the length of the cell cycle. Only the simultaneous analysis of DNA content and BrdUrd incorporation by flow cytometry can measure the rapidity of cell proliferation.

(3) Interphase AgNOR staining and quantification are successfully performed in 100% of the cases examined. This score is never obtained by other methods for cell kinetics evaluation.

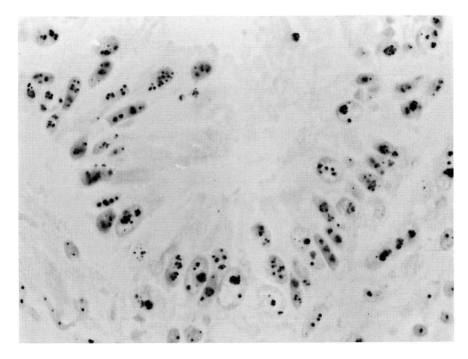

Fig. 7.8. Rapidly proliferating adenocarcinoma of the colon (BrdUrd labelling index, 5.8) stained with the one-step silver method. Note the higher quantity of silver-stained interphase AgNORs in comparison with that of Fig. 7.7. Carnoy fixation. × 360. Reproduced from Derenzini and Trerè (1991a) by permission of Springer–Verlag

DIAGNOSTIC VALUE OF INTERPHASE AgNORs

The analysis of interphase AgNOR distribution in a large series of human cancers (see Table 7.4) has demonstrated that, generally speaking, cancer cells have a greater quantity of interphase AgNORs than the corresponding hyperplastic and normal cells. A malignant lesion can very frequently be distinguished from a benign lesion of the same tissue by simple evaluation of interphase AgNOR quantity.

The importance of a new diagnostic parameter in tumour pathology is strictly related to its capacity to define unambiguously the nature of a tumour lesion. According to Crocker (1990) quantitative or semi-quantitative evaluation of interphase AgNORs can be recommended highly as a reliable diagnostic method in the distinction between benign and malignant neoplasms, or for their grading, only in a few types of lesions. A well-defined difference in interphase AgNOR number was found between benign nevocellular naevi and melanocarcinomas. The various groups (including intradermal, compound, junctional, juvenile and cellular blue naevi) had

Table 7.4. Human tumours in which the interphase AgNOR quantity has been found to have diagnostic relevance

Neoplasia	Reference
Lymphomas	Crocker and Nar (1987) *J. Pathol.*, **151**, 111–118
Cutaneous tumours	Crocker and Skilbeck (1987) *J. Clin. Pathol.*, **40**, 885–889
	Leong and Gilham (1989) *Hum. Pathol.*, **20**, 257–262
	Friedman *et al.* (1991) *Dermatol. Clin.*, **9**, 689–693
Paediatric tumours	Egan *et al.* (1987) *J. Pathol.*, **153**, 275–280
Non Hodgkin's lymphomas	Crocker *et al.* (1988) *J. Pathol.*, **154**, 151–156
Hepatic lesions	Crocker and McGovern (1988) *J. Clin. Pathol.*, **41**, 1044–1048
Salivary gland tumours	Morgan *et al.* (1988) *Histopathology*, **13**, 553–559
	Matsumura *et al.* (1989) *Int. J. Oral Max. Surg.*, **18**, 76–78
	Cardillo (1992) *Acta Cytol.*, **36**, 147–151
Colonic tumours	Derenzini *et al.* (1988) *Virchows Arch. B*, **54**, 334–340
	Yang *et al.* (1990) *J. Clin. Pathol.*, **43**, 235–238
Cervical tumours	Egan *et al.* (1988) *Histopathology*, **13**, 561–567
	Cardillo (1992) *Eur. J. Gynaecol. Oncol.*, **13**, 277–280
Breast tumours	Giri *et al.*, (1989) *J. Pathol.*, **157**, 307–313
	Derenzini *et al.* (1990) *Ulltrastruct. Pathol.*, **14**, 233–245
Rhabdomyoblastic tumours	Eusebi *et al.* (1989) *Tumori*, **75**, 4–7
Mesotheliomas	Derenzini *et al.* (1989) *Acta Cytol.*, **33**, 491–498
Renal tumours	Delahunt *et al.* (1989) *Anal. Cell Pathol.*, **1**, 185–190
	Crocker *et al.* (1990) *Am. J. Clin. Pathol.*, **93**, 555–557
	Bryan *et al.* (1990) *J. Clin. Pathol.*, **43**, 147–148
Bladder tumours	Ooms and Veldhuizen (1989) *Virchows Arch. A*, **414**, 365–369
	Lipponen and Eskelinen (1991) *Anticancer Res.*, **11**, 75–80
Prostatic tumours	Hansen and Ostergard (1990) *Virchows Arch. A*, **417**, 9–13
	Contractor *et al.* (1991) *Urol. Intern.*, **46**, 9–14
	Hansen and Andersen (1992) *APMIS*, **100**, 135–141
Endometrial tumours	Wilkinson *et al.* (1990) *Int. J. Gynecol. Pathol.*, **59**, 55–59
	Papadimitiou *et al.* (1991) *Virchows Arch. A*, **60**, 155–160
Gastric tumours	Rosa *et al.* (1990) *Histopathology*, **16**, 265–269
Bronchial tumours	Abe *et al.* (1991) *Cancer*, **67**, 472–475
Cerebral tumours	Shiraishi *et al.* (1991) *J. Neurosurg.*, **74**, 979–984
	Louis *et al.* (1992) *J. Neuropathol. Exp. Neurol.*, **51**, 150–157
Brain gliomas	Ducrot *et al.* (1991) *Arch. Anal. Cytol. Pathol.*, **39**, 88–93
Meningiomas	Chin and Hinton (1991) *J. Neurosurg.*, **74**, 590–596
Uterine sarcomas	Boquist (1992) *Virchows Arch. A*, **420**, 353–358
Ovarian tumours	Hytiroglou *et al.* (1992) *Cancer*, **69**, 988–992

approximately one interphase AgNOR per cell, whereas lentigo maligna, superficial spreading melanoma and melanocarcinoma had greater interphase AgNOR numbers. There was no overlap between the two types of lesions. Evaluation of interphase AgNOR number also permitted a clear distinction between infiltrating lymphocytes and oat cell carcinoma infiltrates in bronchial material. A very important diagnostic application of the interphase AgNOR parameter was that related to the distribution between neoplastic (both metastatic carcinoma and mesothelioma) cells and reactive mesothelial cells in human pleural effusions (Fig. 7.9). Neoplastic cells had an interphase AgNOR value greater than that of mesothelial reactive cells without overlapping of the mean values. Another kind of tumour in which no overlap was found between benign and malignant interphase AgNOR counts was that of the salivary glands. Finally, it has been shown that interphase AgNOR quantification permitted a total separation between high- and low-grade non-Hodgkin's lymphomas.

With the exception of the above-reported examples, interphase AgNOR quantification cannot be considered to represent a reliable diagnostic parameter to be universally applied in tumour pathology. The method does not permit a malignant cell to be distinguished from the corresponding benign cell on the basis of a higher quantity of interphase AgNORs. This has been clearly demonstrated in breast pathology (Fig. 7.10). Detailed studies carried out in different laboratories on a large number of cases have shown that about 30% of interphase AgNOR values of carcinomas overlap those of benign lesions.

In conclusion, even if in some types of tumour interphase AgNOR quantity has been shown to be of actual diagnostic value, the interphase AgNOR parameter cannot be considered a specific tool for the cytohistological diagnosis of malignancy, very frequently representing only one among the other well-established parameters which help pathologists in tumour diagnosis.

The relationship between interphase AgNOR quantity and cell proliferation rate explains why the AgNOR parameter is not of absolute value for diagnostic purposes. It is in fact well known that the proliferative activity of malignant tumour tissues is not always higher than that of the corresponding hyperplastic and benign lesions.

PROGNOSTIC VALUE OF INTERPHASE AgNORs

The reader of this book has surely perceived that a great effort has been made in the last few years by pathologists in order to characterize better cancer lesions and to define their biological behaviour by using knowledge and methodological approaches derived from basic biological research. Particular attention has been paid to cell kinetics evaluation of cancer lesions.

As already stated, interphase AgNOR quantity does represent a very good

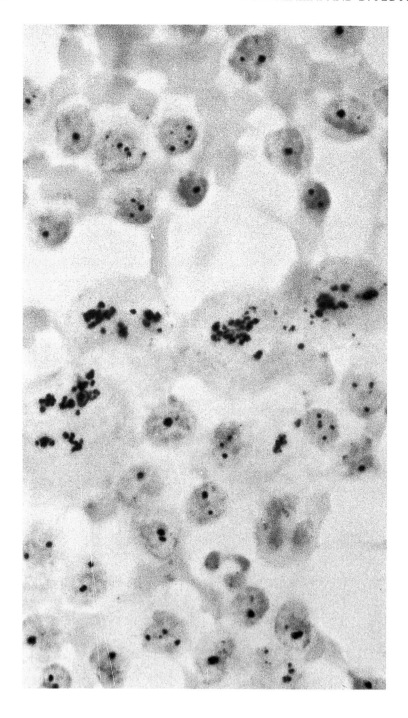

Table 7.5. Human cancers in which the interphase AgNOR quantity has been found to have prognostic relevance

Neuroblastomas	Egan *et al.* (1988) *J. Clin. Pathol.* **41**, 527
Colonic carcinomas	Moran *et al.* (1990) *Br. J. Surg.*, **76**, 1152
	Öfner *et al.* (1990) *J. Pathol.*, **162**, 42
	Rüschoff *et al.* (1990) *Pathol. Res. Pract.*, **186**, 5
Meningiomas	Orita *et al.* (1990) *Neurosurgery*, **26**, 43
Gliomas	Kajiwara *et al.* (1990) *J. Neurosurg.*, **73**, 113
Breast carcinomas	Sivridis and Sims (1990) *J. Clin. Pathol.*, **43**, 390
	Bockmuhl *et al.* (1991) *Pathol. Res. Pract.*, **187**, 437
	Eskelinen *et al.* (1991) *Eur. J. Cancer*, **27**, 989
Prostatic carcinomas	Gillen *et al.* (1988) *Br. J. Surg.*, **75**, 1263
	Lardennois *et al.* (1990) *Acta Urol. Paris*, **24**, 400
	Ploton *et al.* (1992) *Anal. Quant. Cytol. Histol.*, **14**, 14
Sarcomatoid carcinomas of the breast	Eusebi *et al.* (1991) *Ultrastruct. Pathol.*, **15**, 203
Renal cell carcinomas	Delahunt *et al.* (1991) *J. Pathol.*, **163**, 31
Oral squamous cell carcinomas	Sano *et al.* (1991) *J. Pathol. Med.*, **20**, 53
Gastric carcinomas	Kakeji *et al.* (1991) *Cancer Res.*, **51**, 3503
Pharyngeal carcinomas	Pich *et al.* (1991) *Br. J. Cancer*, **64**, 327
Soft tissue sarcomas	Kuratsu *et al.* (1991) *Int. J. Cancer*, **48**, 211
Non-small cell lung carcinomas	Kaneko *et al.* (1991) *Cancer Res.*, **51**, 4008
Mucoepidermoid tumours of salivary glands	Chomette *et al.* (1991) *J. Oral Pathol. Med.*, **20**, 130
Acinic cell carcinomas of salivary glands	Chomette *et al.* (1991) *J. Biol. Buccale*, **19**, 205
Oesophageal carcinomas	Morita *et al.* (1991) *Cancer Res.*, **51**, 5339
Bladder carcinomas	Lipponen and Eskelinen (1991) *Br. J. Cancer*, **11**, 75
	Lipponen and Eskelinen (1992) *Oncology*, **49**, 133
Malignant melanoma	Gambini *et al.* (1992) *Arch. Dermatol.*, **128**, 487
Endometrial carcinomas	Trerè *et al.* (1992) *Virchows Arch. A*, **421**, 203

parameter for cell kinetics evaluation, being strictly related to the rapidity of cell proliferation. The proliferative activity of cancer tissues has important effects on the clinical outcome of patients with neoplastic diseases. It has been shown that measurement of cell kinetics variables offers valuable prognostic indication and may indicate specific treatment to the physician. Despite the fact that evaluation of interphase AgNOR quantity, as a prognostic parameter, has only very recently been introduced into tumour pathology, the relationship between interphase AgNOR quantity and clinical outcome has already been investigated in many types of cancer (Table 7.5). In the majority of the studies a positive correlation between interphase AgNOR amount and survival was

Fig. 7.9 (*opposite page*). Adenocarcinomatous effusion. Four neoplastic cells can be easily distinguished from the reactive non-neoplastic cells on the basis of the AgNOR number. × 1200. Reproduced with permission from *Int. Rev. Exp. Pathol.*, **32**, 1991, Academic Press Inc.

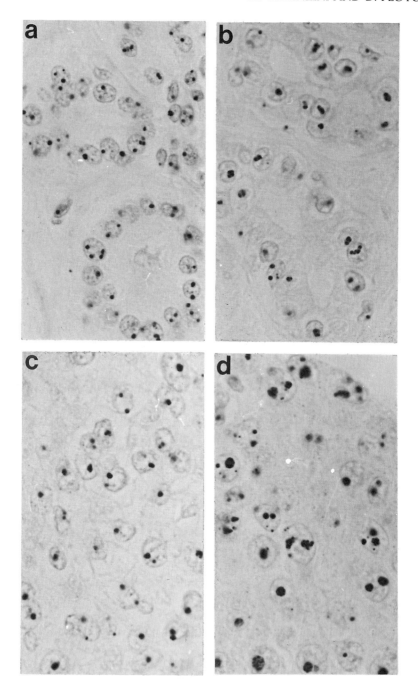

found. The findings regarding the relationship between interphase AgNOR counts and patient survival in colorectal cancers were particularly interesting. Five-year follow-up of patients with colorectal cancers with the same grade of malignancy, determinated according to the established histopathological parameters, showed that tumours of patients with a 5-year survival exhibited at the presentation a significantly lower amount of interphase AgNORs than non-survivors. Moreover, a multivariate survival analysis showed that interphase AgNOR quantity was one of the most important variables predicting death from colorectal carcinoma. Another useful application of the AgNOR parameter for prognostic purposes was found in stage I endometrial adenocarcinomas. In a 10-year follow-up study 90% of patients with low interphase AgNOR values were still alive whereas 60% of patients with high interphase AgNOR scores died of the disease (Figs. 7.11 and 7.12). Once again, in the multivariate analysis of survival, between the interphase AgNOR value and the two most important histological variables—the grade and depth of myometrial invasion—AgNOR quantity appeared to represent the most 'powerful' prognostic parameter.

Fig. 7.10 (*opposite page*). Interphase AgNOR distribution in paraffin-embedded samples of human breast. (a) Normal, (b) adenoma, (c) grade I and (d) grade III ductal infiltrating carcinoma. No differences in AgNOR amounts are detectable in grade I carcinoma and normal or adenomatous cells. × 500. Reproduced with permission from *Int. Rev. Exp. Pathol.*, **32**, 1991, Academic Press Inc.

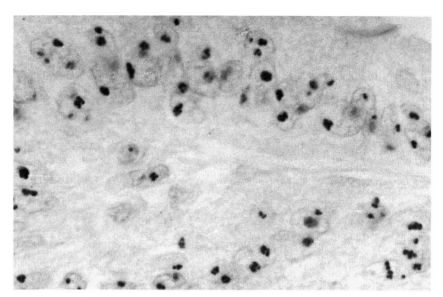

Fig. 7.11. Endometrial adenocarcinoma stained with the one-step silver method. The patient is still alive and well 13 years after diagnosis. Note the low AgNOR amount per cell (mean AgNOR area 2.2 μm^2). × 1140

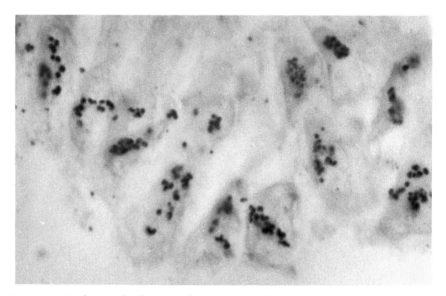

Fig. 7.12. Endometrial adenocarcinoma stained with the one-step silver method. The pa;tient died of the disease 1 year after the surgical resection. Note the high AgNOR amount per cell (mean AgNOR area 5.8 μm^2). × 1140

REFERENCES

Busch H and Smetana K (1970) Nucleoli of tumor cells. In: Busch H and Smetana K (eds), *The Nucleolus*, pp. 448–471. New York: Academic Press.

Crocker J (1990) Nucleolar organizer regions. *Curr. Top. Pathol.*, **82**, 91–149.

Crocker J, Boldy DA and Egan MJ (1989) How should we count AgNORs? Proposals for a standardized approach. *J. Pathol.*, **158**, 185–188.

Derenzini M and Trerè D (1991a) Importance of interphase nucleolar organizer regions in tumor pathology. *Virchows Arch. Cell Pathol. B*, **61**, 1–8.

Derenzini M and Trerè D (1991b) Standardization of interphase Ag-NOR measurement by means of an automated image analysis system using lymphocytes as an internal control. *J. Pathol.*, **165**, 377–342.

Derenzini M, Pession A and Trerè D (1990a) The quantity of nucleolar silver-stained proteins is related to proliferating activity in cancer cells. *Lab. Invest.*, **63**, 137–140.

Derenzini M, Thiry M and Goessens G (1990b) Ultrastructural cytochemnistry of the mammalian cell nucleolus. *J. Histochem. Cytochem.*, **38**, 1237–1256.

Goessens G (1984) Nucleolar structure. *Int. Rev. Cytol.*, **87**, 107–158.

Hadjiolov AA (1985) The nucleolus and the ribosome biogenesis. In: *The Cell Biology Monographs*, pp. 1–267. Vienna: Springer.

Hernandez-Verdun (1983) The nucleolar organizer regions. *Biol. Cell*, **49**, 191–202.

Howell WM (1982) Selecting staining of nucleolus organizer regions (NORs). In: Busch H and Rothblum L (eds), *The Cell Nucleus*, pp. 89–142. New York: Academic Press.

Howell WM and Black DA (1980) Controlled silver-staining of nucleolar organizer regions with a protective colloidal developer: a 1-step method. *Experientia*, **36**, 1014–1015.

Ploton D, Menager M, Jeaesson P *et al.* (1986) Improvement in the staining and in the visualization of the argyrophilic proteins of the nucleolar organizer region at the optical level. *Histochem. J.*, **18**, 5–14.

Ploton D, Visseaux-Coletto B, Canellas JC *et al.* (1992) Semiautomatic quantification of the silver-stained nucleolar organizer regions in tissue sections and cellular smears. *Anal. Quant. Cytol. Histol.*, **14**, 14–23.

Roussel P, Belenguer P, Amalric F and Hernandez-Verdun D (1992) Nucleolin is an AgNOR protein; this property is determined by its amino-terminal domain independently of its phosphorylation state. *Exp. Cell Res.*, **203**, 259–269.

Rüschoff J, Plate KH, Contractor H *et al.* (1990) Evaluation of nucleolus organizer regions (NORs) by automatic image analysis: a contribution to standardization. *J. Pathol.*, **161**, 113–118.

8 Apoptosis: molecular aspects and pathological perspective

M.J. ARENDS and D.J. HARRISON

Apoptosis is a process whereby cells die in a controlled manner, in response to specific stimuli, apparently following an intrinsic programme. Apoptosis occurs often, but not exclusively, in situations to which the term 'programmed cell death' has been applied (Arends and Wyllie, 1991). It is distinct from necrosis where a cell loses its homeostatic control and becomes distended with fluid, leading to lysis and release of intracellular contents with stimulation of inflammation. Apoptosis, by contrast, involves shrinkage and fragmentation of cells with intact membranes and their subsequent removal by phagocytosis before release of harmful intracellular contents can occur.

The significance of apoptosis in many physiological and pathological situations has been increasingly realized since its description 20 years ago (Kerr et al., 1972) (Fig. 8.1). It is a major process determining morphogenesis in the embryo; it is an essential part of the control of cell numbers in tissues that are responsive to alterations in their environment. It is essential for the development of the immune system which is tolerant to 'self' and yet able to respond appropriately to exogenous antigen by selection of the immune repertoire. Induction of apoptosis by cytotoxic T lymphocytes and natural killer (NK) cells is a key effector mechanism in immune defence. Defective entry of lymphocytes into apoptosis appears to be the mechanism of lymphoproliferation and autoimmunity in *lpr* mice that have mutated *fas/APO-1*, a gene whose transmembrane receptor product triggers apoptosis when stimulated. The regulation of acute inflammation at the cellular level is largely determined by the programmed cell death of neutrophils up to 16 h after migrating from the blood vessel to the interstitial site of tissue injury. Elimination of apoptotic neutrophil particles by phagocytosis without release of further inflammatory mediators prevents excessive tissue damage that would follow necrosis and lysis of the cells. Damage following the process of acute inflammation therefore is sustained only for as long as the stimulus persists to recruit fresh neutrophils (Savill et al., 1993).

Much of the pathology seen in a variety of diseases can be attributed to defects in the apoptotic pathway. These defects may occur at a number of

Molecular Biology in Histopathology. Edited by J. Crocker
© 1994 by John Wiley & Sons Ltd

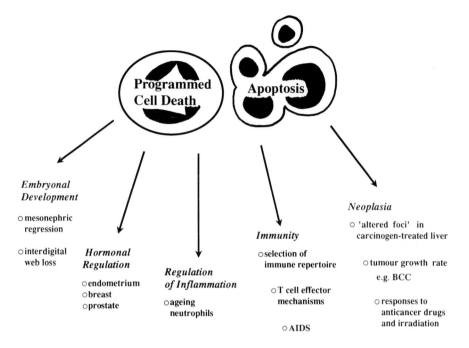

Fig. 8.1. Incidence of programmed cell death/apoptosis in key physiological and pathological situations

levels. Failure to delete autoreactive lymphocytes may result in autoimmunity; prevention of neutrophil apoptosis or induction of neutrophil necrosis/lysis results in excessive tissue destruction (as in staphylococcal pneumonia); aberrant signal transduction of the CD4 receptor on T lymphocytes, caused for example by HIV gp120 protein leads to cell death and, ultimately, the clinical syndrome of AIDS. Dysmorphogenesis and hamartoma formation may reflect abnormalities in programmed cell death mechanisms.

Prevention of apoptosis in tissues, possibly as a result of viral infection or genotoxic injury, may result in excessive cellular proliferation and expansion of a cell population. There is evidence to suggest that such defects underlie carcinogenesis in tissues as diverse as cervix (human papillomavirus (HPV) infection), lymphoma (Epstein–Barr virus (EBV) infection and/or *bcl2* over-expression) and animal models of chemical hepatocarcinogenesis. Net tumour growth rate is markedly influenced by tumour cell loss, of which apoptosis is an important component. Many cancer cells are resistant to the normal induction of apoptosis and failure of tumour cells to undergo apoptosis in response to cytotoxic injury caused by chemotherapeutic drugs probably explains the discrepancy between the known pharmacological mechanisms of tumour drug interactions with their targets and the clinical observations of cancer drug resistance (Hickman, 1992).

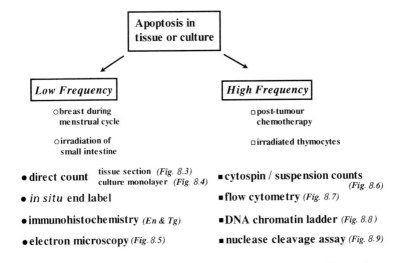

Fig. 8.2. Overview of detection and quantitation of apoptosis according to frequency of occurrence (En = endonuclease; Tg = transglutaminase)

Therefore, apoptosis is of vital importance in a very wide range of physiological and pathological states. With the advent of AIDS and the failure of anticancer therapies to deliver the promise of treatable tumours, there is now increasing interest in the control and mechanisms of apoptosis, its assessment and measurement, and the potential for its manipulation therapeutically.

ASSESSMENT OF APOPTOSIS: MORPHOLOGY AND BIOCHEMISTRY

The detection and quantitation of apoptosis is based on the morphology of the process and/or the underlying biochemical mechanisms involved (Fig. 8.2).

MORPHOLOGY

The morphological changes of apoptosis occur in three phases (Wyllie *et al.*, 1980; Arends and Wyllie, 1991). In the first, there is reduction in both cell and nuclear size, condensation of chromatin into toroids or crescentic caps at the nuclear periphery and nucleolar disintegration with dissociation of the transcriptional complexes from the fibrillar centre. Cells dying by apoptosis detach themselves from their neighbours and from culture substrata. There is loss of specialized surface structures, such as microvilli and contact regions, such that the cell adopts a smooth contour. Cell volume shrinks, cytoplasmic organelles become compacted, and the smooth endoplasmic reticulum dilates. The dilated cisternae fuse with the cell membrane, giving rise to a bubbling appearance at the surface. Cytoskeletal filaments aggregate in side-to-side arrays, often parallel to the cell surface, and ribosomal particles clump in semi-crystalline formations, but otherwise the organelles remain intact. In contrast to necrosis—the other major type of cell death—mitochondria do not show 'high-amplitude swelling', the cell membrane does not become permeable to vital dyes at this stage, and apoptotic cells within tissues do not elicit an acute inflammatory reaction.

In phase 2 (which may overlap with the first), there is a blebbing at the cell surface and crenation of the nuclear outline. Both nucleus and cytoplasm may split into fragments of various sizes. Typically, the cell becomes a cluster of round, smooth, membrane-bounded 'apoptotic bodies', some containing nuclear fragments, and others without. These bodies may be shed from epithelial surfaces or phagocytosed by neighbouring cells or macrophages.

In phase 3, there is progressive degeneration of residual nuclear and cytoplasmic structures. In cultured cells, this is manifested as membrane rupture producing permeability to vital dyes. In tissues these changes (sometimes termed 'secondary necrosis') usually occur within the phagosome of the ingesting cell. Eventually membranes disappear, organelles become unrecognizable and the appearance is that of a lysosomal residual body. The majority of apoptotic bodies seen in tissues studied with the light microscope are in this phase, and sometimes the smooth outline of the ingesting phagosome can be seen around them, but earlier phases can also be recognized by their rounded contours and deeply hyperchromatic, often fragmented, nuclei set in markedly eosinophilic cytoplasm.

Time-lapse cinematographic studies of apoptosis (Evan et al., 1992) reveal the sudden onset of cell separation from neighbours, net shrinkage, and surface blebbing and bubbling, as cells enter phase 1 and 2, after a variable time from exposure to the lethal stimulus. This initial response lasts for only a few minutes, and generates small, dense apoptotic cells. If not phagocytosed immediately these cellular particles undergo a gradual loss of cell density, coinciding with loss of membrane integrity, shown ultrastructurally and by failure to exclude vital dyes. Apoptotic cells remain recognizable within tissues for 3–6 h, a time-course which coincides with that of complete degradation of other large biological structures within the phagosomes of macrophages. This relatively short period ensures that high rates of apoptosis produce only small increases in the proportion of apoptotic cells observed in tissue sections.

CHROMATIN CLEAVAGE

Internucleosomal chromatin cleavage is associated almost exclusively with the morphology of apoptosis. This association was first demonstrated in glycocorticoid-treated thymocytes (Wyllie, 1980). Cleavage of internucleosomal linker DNA generates well-organized chains of oligonucleosomes, with DNA lengths that are integer multiples of 180–200 bp—the size of DNA wrapped around a single histone octamer—observed as a 'chromatin ladder' on gel electrophoresis. Chromatin laddering has now been reported along with morphological chromatin condensation of apoptosis in many cell systems. Chromatin cleavage products of 50 kb and 300 kb DNA fragments, which may represent supercoiled chromatin loops, may also be found in some examples of morphologically apoptotic cells.

It has recently been demonstrated that DNA cleavage in apoptosis occurs selectively, without associated chromatin proteolysis. The nuclear matrix appears normal, in terms of structural organization and the presence of the most abundant protein species. DNA cleavage is at widely dispersed sites: the apoptotic nucleus has a normal content of acid-precipitable DNA. Two classes of oligonucleosome chromatin fragments are generated: 70% exists as oligonucleosome fragments bound to the nucleus, while 30% is unattached. Although the bound chromatin includes fragments as short as dinucleosomes, the majority are long; in contrast, the unattached chromatin comprises mono- and short oligonucleosome fragments. This minority class probably derives from chromatin in a transcriptionally active configuration as the chromatin-bound proteins are depleted of histone H1 and enriched in high-mobility group (HMG) proteins 1 and 2—changes associated with active gene transcription. Whereas inactive heterochromatin is thought to be tightly wound in a solenoid, transcriptionally active chromatin is not compacted in this way, which would allow better access to enzymes in the nucleoplasm, producing more complete digestion. The pattern of chromatin digestion in apoptosis, therefore, is consistent with activation of an endonuclease in solution in the nucleoplasm, rather than a constituent of the matrix itself. The characteristic nucleolar morphology in apoptosis can also be explained in terms of cleavage of the transcriptionally active ribosomal genes, with conservation of the nucleolin-rich fibrillar centre. The chromatin cleavage, nucleolar morphological changes and chromatin condensation can be closely mimicked by micrococcal nuclease digestion of normal thymocyte nuclei in the presence of protease inhibitors. Thus, in apoptosis, selective activation of an endogenous endonuclease appears to be responsible not only for widespread chromatin cleavage but also the major nuclear morphological changes (Arends et al., 1990).

Early experiments with thymocyte nuclei suggested that they contained an enzyme capable of cleaving chromatin in apoptosis. If incubated in neutral pH, together with both calcium and magnesium, such nuclei quickly developed multiple double-strand DNA breaks, generating the familiar 'ladder' on electrophoresis (Cohen and Duke, 1984). This activity can be inhibited by zinc ions. The thymocyte nuclease activity can be eluted from nuclei and used to cleave chromatin of a target system (nuclei from cells labelled during growth with tritiated thymidine) to release labelled oligonucleosomes. This nuclease activity is optimum at pH 7.5, in contrast to contaminating acid nucleases, which also differ in cleaving DNA to much smaller (acid-soluble) fragments. The neutral calcium–magnesium-sensitive endonuclease is maximally eluted from normal thymocyte nuclei at 300 mM NaCl, and appears to be an anionic protein of molecular weight greater than 110 kDa (Wyllie et al., 1992). In certain cell lines undergoing apoptosis in vitro in response to glucocorticoid, the extractable calcium–magnesium-sensitive endonuclease activity rose from low levels, peaking as endogenous chromatin cleavage and the morphology of apoptosis appears (Wyllie et al., 1992). Other candidate nucleases have been suggested but no definitive characterization has yet emerged.

TRANSGLUTAMINASE ACTIVATION

Induction and activation of tissue transglutaminase (Ca^{2+}-dependent protein–glutamine γ-glutamyltransferase) has been associated with apoptosis during involution of liver hyperplasia and in glucocorticoid-treated thymocytes. Transglutaminases cross-link proteins through ε-(γ-glutamyl) lysine bonds and mediate formation of cornified envelopes by epidermal keratinocytes, cross-linking of fibrin and 2-plasmin inhibitor in the final stages of thrombus stabilization. In apoptosis there is an increase in transglutaminase mRNA and protein, enzyme activity and protein-bound (γ-glutamyl) lysine. The probable consequence of transglutaminase activation is an extensive cross-linking of cytoplasmic and membrane proteins. Apoptotic cells have been shown to contain protein shells insoluble in detergents and chaotropic agents. These shells, which are not extractable from normal cells, appear in scanning electron microscopy as wrinkled, spherical structures with some morphological similarities to epidermal cornified envelopes. This action may limit escape of potentially toxic intracellular contents (e.g. lysosomal enzymes) that may otherwise excite an inflammatory response.

Intracellular signalling is also important in apoptosis and is mediated, at least in part, by various isoforms of protein kinase C.

DETECTION AND MEASUREMENT OF APOPTOSIS

In situations where the expected prevalence of apoptosis is relatively high (greater than 10%), for example after cytotoxic drug treatment of a tumour or irradiation of thymocytes in vitro, the detection of apoptosis is readily accomplished. However, the rate of apoptosis, particularly in situations encountered in vivo, is often much lower. This is partly because of the very short time, less than 6 h, that apoptotic bodies remain identifiable in tissue sections. However, even apparently low prevalences of apoptosis (less than 2%) can have profound effects over time in tissues. For these reasons the most appropriate methodologies used to detect and quantify apoptosis vary according to the extent of apoptosis likely to be encountered. These are listed in Fig. 8.2.

The most commonly used method is direct identification of cells with

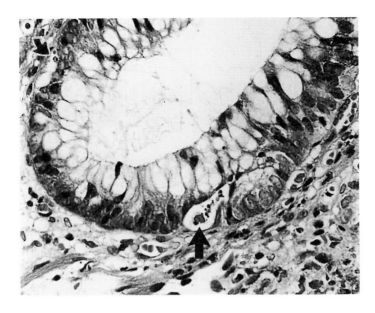

Fig. 8.3. Apoptosis seen in a tissue section. This is HIV 'cryptitis' in a rectal biopsy from a patient with HIV1 infection and AIDS. Note the classical fragmented apoptotic cell (large arrow) in an area of vacuolation associated with a lymphocyte which has infiltrated the crypt. A smaller apoptotic body is seen at another part of the crypt (arrow, upper left)

recognizable nuclear changes of apoptosis by light microscopy in tissue sections (Fig. 8.3) or refractile, shrunken cells in tissue culture (Fig. 8.4). Counting the percentage of apoptotic cells has been greatly facilitated by interactive computerized systems such as the HOME microscope. These allow large numbers of cells to be counted accurately. The specificity of the morphological changes seen in tissue sections or culture can be confirmed by electron microscopy (Fig. 8.5). As will be clear from the above, apoptosis can only be recognized by light microscopy at a stage where there is obvious shrinkage or fragmentation of the nucleus. Therefore, the count tends to be an underestimate and this can result in inaccuracy when only small numbers of events are detected. There are several strategies to attempt to overcome this problem. Since apoptosis involves the fragmentation of chromatin, several groups have used DNA polymerases or terminal transferases to end-label DNA strand breaks by the incorporation of biotinylated nucleotides—*in situ* end-labelling (ISEL). Labelled nuclei are then identified by addition of a streptavidin–peroxidase conjugate and an appropriate peroxidase substrate (Wijsman *et al.*, 1993). These methods have several potential advantages: more nuclei are detected as being apoptotic; there is less equivocation since apoptotic nuclei are clearly decorated; and the labelled nuclei in tissue sections

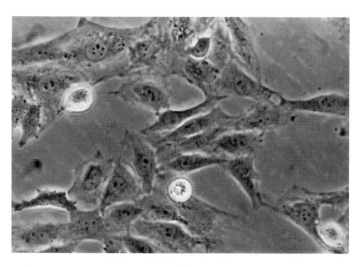

Fig. 8.4. Apoptosis in a monolayer of fibroblasts in tissue culture. The apoptotic cells are shrunken, spherical and refractile bodies that float on top of the adherent cells

may be counted by automated image cytometry. Disadvantages include the small window of enzymic activity in which apoptotic but not non-apoptotic cells are labelled. Another strategy seeks the same advantages by immunological detection of proteins, such as endonuclease or transglutaminase, which are essential components of the apoptotic mechanism. To date this strategy has been successful with transglutaminase detection, but remains more theoretical with regard to endonuclease detection, in part because extraction and purification of the endonuclease(s) involved has been elusive.

Where apoptosis is present at a high frequency, direct counts of sections, cytospins or suspensions can be used (Fig. 8.6). Several groups have developed techniques to assess apoptosis semi-quantitatively by flow cytometry (Darzynkiewicz *et al.*, 1992). We have used acridine orange to stain nuclei for flow cytometry. Apoptotic cells are smaller and less fluorescent than viable cells (Fig. 8.7). Necrotic cells are not assessed because they do not fluoresce green with acridine orange since their DNA is denatured and loses its double-stranded configuration. The 'gold standard' for apoptosis has been long regarded by the production of a 'chromatin ladder' by gel electrophoreses (Fig. 8.8) (Wyllie, 1980), although this is not always found in a number of cell lines (e.g. MCF 7) even when apoptosis is clearly identifiable on morphological grounds. Enrichment of apoptotic cells prior to electrophoresis can be achieved by centrifugation of cells over a Percoll gradient. Since chromatin cleavage occurs by the action of an endonuclease a further detection method relies on the ability of this enzyme in a crude cell extract to cleave plasmid DNA (Wyllie *et al.*, 1992) (Fig. 8.9).

It is worth stressing the importance of the qauntitation of apoptosis, rather than simply its detection. For example, the cell count in the lymph nodes of a mouse is reduced by 50% in 48 h following intravenous injection of anti-CD4 antibody. This loss of cells is accounted for by apoptosis. The maximum prevalence of apoptosis in sections of lymph nodes, at only 1.7%, is 1.6% higher than background apoptosis in unstimulated lymph nodes from untreated animals. Thus, very small perturbations in the rate of apoptosis have profound effects on total cell population.

RECOGNITION AND ELIMINATION OF APOPTOTIC BODIES: RELEVANCE TO THE CONTROL OF INFLAMMATION

The elimination of apoptotic cells is critical if an inflammatory response and subsequent damage to neighbouring cells is to be avoided following cell death. This has been demonstrated in the nematode *Caenorhabditis elegans* (Hengartner *et al.*, 1992) and morphogenesis of *Drosophila* spp. It is clear that although macrophages are of great importance for the elimination of apoptotic cells, other cells types, sometimes regarded as facultative phagocytes, can also be involved. In human biology perhaps the best example demonstrating the importance of elimination of apoptotic cells is that of regulation of acute inflammation. Neutrophils are attracted to sites of inflammation, where their primary role is phagocytosis, killing and degradation of bacteria or damaged cells. However, should neutrophils have a prolonged life span they would continue to function after the initial stimulus had ceased. Neutrophils have a short half-life once they have left the circulation and this appears to be of major significance in the control of inflammation. If neutrophils died by necrosis, cell membranes would lyse and the release of lysosomal contents would cause extensive local tissue damage, eliciting a further inflammatory response. However, there is recent evidence that neutrophils are, in fact, removed by undergoing a process of apoptosis and subsequent phagocytosis by macrophages (Savill *et al.*, 1993). This mechanism ensures that neutrophil contents are not released, and there is no triggering of macrophage cytokine production. It has also been suggested that defective elimination of apoptotic bodies may, of itself, be responsible for some forms of inflammatory disease, such as glomerulopathies and the polyclonal B cell activation seen in systemic lupus erythematosus.

In studies of acute inflammation, Savill and colleagues have described three possible recognition mechanisms. First, apoptotic cells appear to undergo changes in surface carbohydrate moieties. This may be due to loss of sialic acid residues from surface glycoproteins allowing the apoptotic cell membrane to be recognized by specific phagocyte lectin-like receptors. Second, thrombospondin associates with a thrombospondin-binding moiety on the surface of the apoptotic cell and is recognized by thrombospondin receptors on the phagocyte. These are thought to comprise CD36 and $\alpha_V\beta_3$ vitronectin

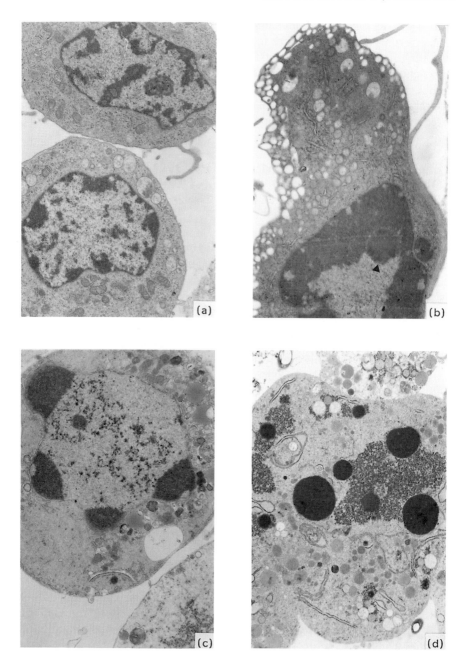

receptor. The third potential mechanism is alteration of the distribution of membrane phospholipids, in particular phosphatidylserine. Once recognition and phagocytosis have occurred there is a breakdown of the phagocytosed apoptotic body within the macrophage (Fig. 5e and f). Macrophages do not secrete cytokines in response to phagocytosis of apoptotic neutrophils, possibly because of the Fc-independent nature of the phagocytosis. However, it is not yet clear whether the recognition and phagocytosis of apoptotic bodies in this way represent a negative feedback loop for the inflammatory process.

PRIMING AND TRIGGERING ARE REQUIRED FOR APOPTOSIS

Apoptosis depends upon the availability of a number of key proteins, including calcium–magnesium-sensitive endonuclease, γ-glutamyl transpeptidase and

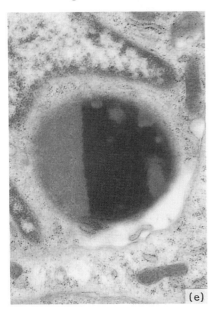
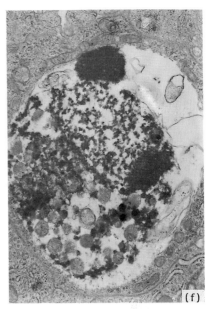

Fig. 8.5. Apoptosis in electron micrographs. (a) Viable fibroblasts from tissue culture showing mixed heterochromatin and euchromatin within nuclei. (b) Early-phase apoptosis with condensation of chromatin around the periphery of the nucleus and a nucleolar fibrillar centre (arrowhead) closely apposed to chromatin. Dilated endoplasmic reticulum vacuoles are present below the plasma membrane. (c) Aggregation of condensed chromatin into masses at the nuclear periphery. Nucleolar disintegration of granular and dense fibrillar components is evident. (d) Multiple spheres of condensed chromatin are present within a mid-phase apoptotic cell, together with semi-crystalline arrays of ribosomal particles. (e) Phagocytosed apoptotic body showing condensed chromatin. (f) Secondary necrosis of apoptotic body. Remnants of condensed chromatin are still visible

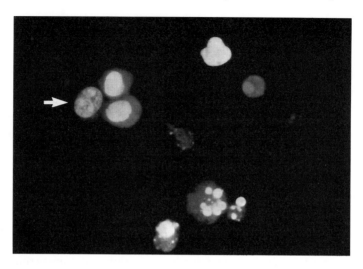

Fig. 8.6. Apoptotic cells in suspension identified by acridine orange staining and examination under ultraviolet light. A normal cell, with nucleus heterogeneously stained, is identified by an arrow. Adjacent to this are two early-phase apoptotic cells showing condensed, homogeneous chromatin. Elsewhere mid- and late-phase apoptotic cells with multiple spheres of condensed chromatin are seen

many other unknown effectors. None of these molecules is present in every cell of a tissue. Therefore they must accumulate, at least in some cells, before apoptosis can occur. In some circumstances this may occur immediately before the onset of apoptosis. Their coordinate expression, regulated by presumed controller genes, is therefore essential priming for apoptosis. In some tissues, for example the thymic cortex, there is a high proportion of primed cells but in most tissues, for example liver, primed cells are a small minority. The initiation of apoptosis in primed cells is the result of a distinct set of events and is known as triggering. A number of triggers have been described, including the controlled influx of calcium into the cell, causing activation of endonuclease and protein kinases. It is probable that in many tissues priming is a reversible or inhibitable state, but it is likely that once triggered a cell proceeds inevitably into apoptosis. In many cells there is a shut-down of total protein and RNA synthesis early in apoptosis but in other cell types the initiation of apoptosis is dependent upon protein synthesis and its triggering can be abrogated by application of inhibitors such as cycloheximide or actinomycin D, shortly after the lethal stimulus.

The genetic regulation of programmed cell death in nematodes has been well characterized in studies of *C. elegans*. Analysis of mutations yielding abnormal cell death phenotypes identified three loci of particular importance, named *ced-3*, *ced-4* and *ced-9*. Recessive mutations at the *ced-3* and *ced-4* loci block nearly all of the programmed cell death that occurs during development

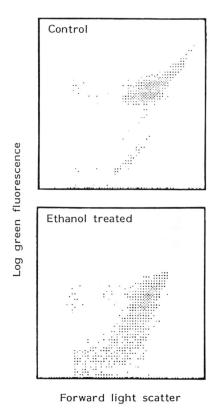

Fig. 8.7. Flow cytometric analysis of apoptosis. H9 human leukaemia cells have been stained with acridine orange before (control) and after treatment with 3% ethanol. The axes are log green fluorescence (representing double-stranded DNA content) and forward angle light scatter (representing cell size). Control cells are fairly uniform in size and DNA content, as expected. On induction of apoptosis by ethanol, cells shrink and fragment, thereby reducing DNA content (green fluorescence) and becoming smaller (forward scatter). Cells on the x axis with no green fluorescence represent secondary necrosis with denaturation of double-stranded DNA

of *C. elegans* (Hengartner *et al.*, 1992). It has been suggested that these genes code for specific functions within the dying cell. Some loss-of-function mutations of the *ced-9* gene, with recessive transmissibility, exhibit an embryonic lethal phenotype, consistent with constitutively activated cell death. Gain-of-function mutations of *ced-9* exhibiting dominant transmissibility are phenotypically similar to *ced-3* and *ced-4* mutations. These observations suggest that the *ced-9* product may act as a type of master-switch able to regulate positively or negatively the cell death pathway. These studies have already shed much light on the type of genes regulating cell death in general, including mammalian apoptosis. Despite the species divergence, many of the

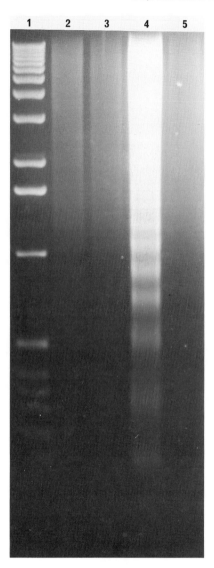

Fig. 8.8. Oligonucleosomal-sized DNA fragments characteristic of the 'chromatin ladder' of apoptosis. 1.5% agarose gel electrophoresis of DNA extracted from cellular bodies floating within culture medium above monolayers of Rat-1 fibroblasts constitutively expressing either functionally active c-myc (Δ 145–262) (lanes 2 and 4) or inactivated c-myc (Δ 106–144) as a control (lanes 3 and 5). All cultures were grown in 0.1% fetal calf serum, with supernatant cells harvested at time 0 h (lanes 2 and 3) and time 30 h (lanes 4 and 5). A 1 kb ladder size marker was included (lane 1) (experiment fully described by Evan *et al.*, 1992). Fibroblasts expressing functionally active c-myc under conditions of serum deprivation displayed high levels of apoptosis shown by the 'chromatin ladder' (lane 4)

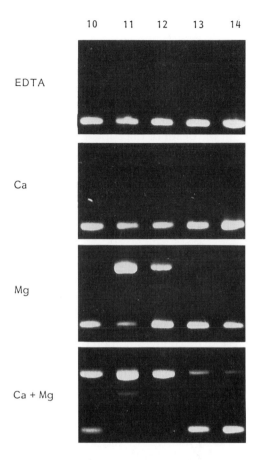

Fig. 8.9. Plasmid cleavage assay for nuclease activity. The electrophoretic mobility of supercoiled, circular plasmid DNA is altered by single-strand nicks (relaxes the supercoiled circles) and double-strand cleavage (produces linear molecules), permitting the sensitive identification of nuclease activity extracted from viable rat thymocyte nuclei (Wyllie *et al.*, 1992). Plasmid DNA was incubated with successive fractions (10–14) of protein from superose 12 gradient sizing columns, in the presence of 2 mM EDTA, calcium, magnesium, or both cations together, and visualized by 1% agarose gel electrophoresis. The uncut (supercoiled) form remained after incubation for 10 h at 37 °C in the presence of EDTA or calcium. Nicking activity, peaking in fractions 11 and 12 (corresponding to 110–130 kDa molecular weight protein complexes), was seen after incubation with magnesium and enhanced (with some double-strand cleavage) by both cations together. Reproduced from Wyllie *et al.* (1992) by permission of Academic Press Ltd

structural features of programmed cell death in the nematode are similar to those of apoptosis. To date, however, mammalian homologues closely related to *ced-3*, *ced-4* or *ced-9* have not been identified, although there is partial similarity, both structurally and functionally, between *bcl-2* and *ced-9*.

GENETIC REGULATION OF APOPTOSIS IN TUMOURS

Conventionally, tumours have been regarded as masses of tissue with an increased proliferation rate. Although this may often be the case it is apparent that many tumours with high levels of cell division, in fact, grow slowly (e.g. basal cell carcinoma of skin). Furthermore, some tumours with low proliferation rates (e.g. low-grade lymphoma or myeloma) inexorably increase in bulk, eventually resulting in the death of the patient. Therefore, control of tumour cell population expansion—i.e. the rate of increase in numbers of cells—is not merely a reflection of the proliferation rate, but a balance of cell loss and cell gain. It has long been known that many tumours have high rates of apoptosis and may lose over 90% of their cells by this route.

For several reasons, there is currently much interest in the genes that regulate apoptosis in tumours. First, some oncogenes, such as activated Kirsten *ras*, are directly permissive for population expansion, and as part of this action appear to inhibit cell entry into apoptotic pathways as well as to stimulate cell proliferation. Second, mutation of certain tumour suppressor genes, such as *p53*, may also have the dual effect of promoting proliferation and inhibiting entry to apoptosis. Third, most chemotherapeutic and radiotherapeutic regimes for the treatment of cancer induce apoptosis in susceptible cells (Hickman, 1992). Despite much investigation of the pharmacological basis of tumour drug resistance it is clear that this critically important clinical phenomenon cannot be explained solely on pharmacological grounds. There is increasing evidence that at least one mechanism of drug resistance may be failure of cells to respond in an appropriate way, i.e. to enter apoptosis, after sustaining injury. An understanding of the genetic control of apoptosis and the prevention of apoptosis is therefore of likely benefit to our whole understanding of tumorigenesis and response to therapy. Since histology is often used in the assessment of the prognosis of tumours, the prospect of being able to comment on the extent of apoptosis as well as mitosis in tumours has implications for practitioners of histopathology.

In Fig. 8.10 there is an attempt to represent schematically the sites of action or control of a number of known oncogenes and tumour-suppressor genes in the growth of human tumours. Both low- and high-turnover populations can result in population expansion, although it is apparent that many tumours with a high turnover are also particularly susceptible to apoptosis. The *ras* oncogene influences tumour growth in many circumstances; for example, it is mutated in a substantial proportion of colorectal adenomas, including very small adenomas and even a proportion of hyperplastic/metaplastic polyps not generally considered to be neoplastic. Greater than 90% of pancreatic carcinomas have a mutation in the Kirsten *ras* oncogene. Transgenic animal models of aberrant *ras* expression in pancreas also confirm that *ras* is associated with excessive cell population expansion in the pancreas, and these animals rapidly develop pancreatic carcinoma. The *ras* protein product p21 is

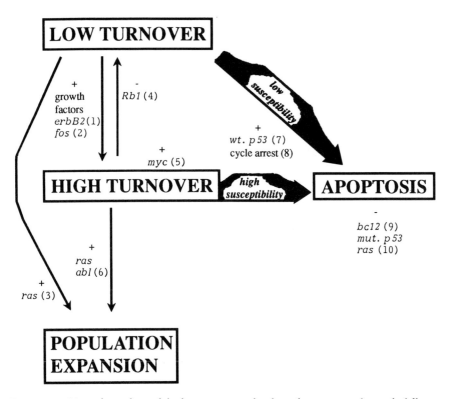

Fig. 8.10. Hypothetical model that compounds data from a number of different circumstances, to illustrate potential regulation by certain genes of transit between tumorigenic states of low turnover, high turnover of cells (high proliferation and high apoptosis) and population expansion. Some genes may act at other points as well as their key sites of action shown here. Other genes, not shown, may also act at these transition control points. *Examples:* 1. Breast cancer. 2. 'Altered foci' in experimental hepatocarcinogenesis; 3. Colonic adenoma; 4. Retinoblastoma; 5. Burkitt's lymphoma; 6. Chronic myeloid leukaemia; 7. Genotoxic damage in thymocytes; 8. Cytotoxic drugs, e.g. etoposide; 9. Low-grade lymphoma; 10. Pancreatic carcinoma.   *Notes:* A. Genes often act in conjunction with one another, rather than having a single attributable role; B. Some genes (e.g. Rbl), may have a window of opportunity to fulfill their postulated role.

associated with signal transduction at the inner surface of the cytoplasmic membrane and mutation effectively results in p21^{ras} being constitutively switched 'on'. Thus, the cell responds as if it had been continuously stimulated by growth factors. There is also evidence from *ras*-transformed cell lines that *ras* can inhibit the apoptotic pathway as well as stimulate proliferation, resulting in marked cell population expansion and tumour growth (Arends *et al.*, 1993).

The change from a low turnover to a high turnover growth profile in tissues is controlled by a number of oncogenes. Growth factors and their pathways can bring about this transition. In experimental chemical hepatocarcinogenesis proliferative foci in the liver appear in association with up-regulation of the c-fos oncogene. C-erbB2 is amplified in many invasive and *in situ* breast cancers. The maintenance of a high turnover state (for example in lymphoma) is often associated with constitutive over-expression or deregulation of c-myc. As well as being permissive for proliferation, expression of the c-myc oncogene increases susceptibility to apoptosis (Evan et al., 1992). This is particularly true if there is cell cycle arrest, caused by growth factor deprivation, selective nutrient deprivation or treatment with certain chemotherapeutic drugs. In normal (non-neoplastic) cells, cell cycle arrest induces down-regulation of c-myc and this subsequently leads to entry into a stable state of growth arrest or sometimes differentiation. Thus, myc appears to be the central regulator of cell fate, its down-regulation influencing growth arrest and differentiation, and its constitutive expression inducing proliferation and high susceptibility to apoptosis. However, there appears to be a complex network of interacting genes that may override or modify these myc effects.

Bcl-2 can inhibit entry to apoptosis from low- or high-turnover states and therefore produce increased population expansion (Hockenbery et al., 1990). Bcl-2 is believed to be the major cause of gradual population expansion in low-grade lymphoma that typically has a prolonged indolent course but is virtually refractory to chemotherapy. This is consistent with the failure to enter apoptosis when chemotherapeutic triggers are applied. The abl oncogene also appears to drive population expansion directly through bypassing entry into apoptosis, and this is associated with the marked population expansion seen in chronic myeloid leukaemia, although the relatively mature myeloid cells produced still retain the ability to enter apoptosis.

The oncosuppressor retinoblastoma gene (Rb) plays a role in genetic regulation of apoptosis, at least in a critical window of opportunity in embryonic development, as shown in experimentally gene-targeted mice homozygous for Rb deletions that die *in utero*, in which there is malformation of the central nervous system (CNS) with very prominent increase in apoptosis in several CNS structures, including spinal ganglia (Clarke et al., 1992). Inactivation of both Rb alleles is essential for development of both sporadic and hereditary retinoblastomas.

Experiments on thymocytes taken from gene-targeted mice homozygous for p53 deletions showed normal induction of apoptosis by physiological stimuli such as glucocorticoid, but absence of the expected rise in apoptosis after treatment with genotoxic agents such as irradiation or etoposide (Clarke et al., 1993). Mice heterozygous for the p53 deletion showed an intermediate apoptotic response, indicating a gene dosage-dependent effect. This is consistent with the previously suggested p53 role of guardian of the genome against genotoxic damage, and suggests that p53 mediates this protective function either through induction of G_1 arrest to allow DNA repair or via initiation of apoptosis, perhaps if damage is perceived to be too severe. The *Val-135* mutant of *p53* behaves like other oncogenic *p53* mutants at 37.5 °C, but like wild-type *p53* at 32.5 °C. Following its introduction into a leukaemic cell line, wild-type *p53* activity resulted in induction of apoptosis, not found with mutant *p53* activity. Thus, many genes that are altered in cancers appear to influence both proliferation and apoptosis, affecting the net tumour growth rate and response to therapy.

CONCLUDING REMARKS

Apoptosis has very extensive and important biological functions. It occurs in a variety of physiological and pathophysiological situations. It is one of the major mechanisms in the development of a lymphocyte repertoire in the immune system as well as being an end-effector mechanism for the immune response. It is an integral part of the regulation of tissue homeostatis and its deregulation can lead to conditions as disparate as autoimmune disease, AIDS, tumorigenesis and acquired resistance of tumour cells to chemotherapy. There have been many advances in our understanding of the triggering mechanisms of apoptosis, the biochemistry of the changes that ensue and the genetic regulation of entry into apoptosis. The histopathologist has the potential to add to these studies, in many of the situations mentioned, by identifying and quantifying apoptosis within tissues and cell populations. Since very small perturbations of the apoptotic rate can result in quite excessive depletion or expansion of cell populations, quantitation of this phenomenon is largely, and will probably remain, within the province of the practising histopathologist.

ACKNOWLEDGEMENTS

The work described here has been supported by the Cancer Research Campaign, the Scottish Hospital Endowments Research Trust, the Scottish Home and Health Department and the Medical Research Council. We are grateful to Professor Andrew Wyllie for helpful advice.

REFERENCES

Arends MJ, Morris RG and Wyllie AH (1990) Apoptosis: the role of the endonuclease. *Am. J. Pathol.*, **136**, 593–608.

Arends MJ and Wyllie AH (1991). Apoptosis: mechanisms and roles in pathology. *Int. Rev. Exp. Pathol.*, **32**, 223–254.

Arends MJ, McGregor AH, Taft NJ, Brown EJH, Wyllie AH (1993) Susceptibility to apoptosis is differentially regulated by *c-myc* and mutated *Ha-ras* oncogenes and is associated with endonuclease availability. *Br. J. Cancer*, **68**, 1127–1133.

Clarke AR, Maandag ER, van Roon M *et al.* (1992) Requirement for a functional Rb-1 gene in murine development. *Nature*, **359**, 328–330.

Clarke AR, Purdie CA, Harrison DJ *et al.* (1993) Thymocytes apoptosis induced by p53: dependent and independent pathways. *Nature*, **362**, 849–852.

Cohen JJ and Duke RC (1984) Glucocorticoid activation of a calcium dependent endonuclease in thymocyte nuclei leads to cell death. *J. Immunol.*, **132**, 28–42.

Darzynkiewicz Z, Bruno S, Del Bino G *et al.* (1992) Features of apoptotic cells measured by flow cytometry. *Cytometry*, **13**, 705–808.

Evan G, Wyllie A, Gilbert C *et al.* (1992) Induction of apoptosis in fibroblasts by *c-myc* protein. *Cell*, **69**, 119–128.

Hengartner MO, Ellis RE and Horvitz HR (1992) *Caenorhabditis elegans* gene ced-9 protects cells from programmed cell death. *Nature*, **356**, 494–499.

Hickman JA (1992) Apoptosis induced by anticancer drugs. *Cancer Metastasis Rev.*, **11**, 121–139.

Hockenbery D, Nunez G, Milliman C, Schreiber RD and Korsmeyer SJ (1990) Bcl-2 is an inner mitochondrial membrane protein that blocks programme cell death. *Nature*, **348**, 334–336.

Kerr JFR, Wyllie AH and Currie AR (1972) Apoptosis: a basic biological phenomenon with wide-ranging implications in tissue kinetics. *Br. J. Cancer*, **26**, 239–257.

Savill JS, Fadok V, Henson PM and Haslett C (1993) Phacogyte recognition of cells undergoing apoptosis. *Immunol. Today*, **14**, 131–136.

Wijsman JH, Jonker RR, Keijzer R *et al.* (1993) A new method to detect apoptosis in paraffin sections: in situ end labelling of fragmented DNA. *J. Histochem. Cytochem.*, **41**, 7–12.

Wyllie AH (1980) Glucocorticoid induced thymocyte apoptosis is asociated with endogenous endonuclease activation. *Nature*, **284**, 555–556.

Wyllie AH, Kerr JFR and Currie AR (1980) Cell death: the significance of apoptosis. *Int. Rev. Cytol.*, **68**, 251–306.

Wyllie AH, Arends MJ, Morris RG, Walker SW and Evan G (1992) The apoptosis endonuclease and its regulation. *Semin. Immunol.*, **4**, 389–397.

Index

Acrocentric chromosomes, 136
Actinomycin D, 162
Acute lymphocytic leukaemia (ALL), 27
Acute myeloid leukaemia (AML), 27, 67
Acyclovir, 13
AgNOR
  enumeration, 101
  interphase, 122, 132
    and patient survival, 147
    diagnostic value, 141–2
    distribution in cancer cells, 136–8
    in tumour pathology, 135–47
    prognostic value, 143–7
    quantity as parameter of cell
      kinetics, 138–40
  proteins, 121, 125, 127, 131, 133,
    137–8
  quantification techniques, 133–5
  scores, 119
  staining method, 132
AIDS, 152, 153, 157
Allele-specific oligonucleotide (ASO)
  probes, 31
Aneuploidy
  associated with congenital
    abnormality, 66
  associated with haematological
    malignancy, 67, 68
  detection, 62, 64–5
Anti-oncogenes, PCR, 52–3
Apoptosis, 151–70
  assessment of, 153–6
  detection and measurement of, 156–9
  electron microscopy, 157
  flow cytometry, 163
  frequency of occurrence, 153
  genetic regulation in tumours, 166–9
  gold standard, 158
  high frequency, 158
  in disease pathology, 151–2
  in monolayer of fibroblasts in tissue
    culture, 158
  incidence, 151

morphology, 153–4
overview of detection and
  quantitation, 153
prevention of, 152
priming and triggering requirements
  for, 161–5
process of, 151
quantitation, 158–9
significance of, 151
time-lapse cinematographic studies,
  154
Apoptotic cells, 156, 158
  in suspension, 162
  recognition and elimination of, 159–61
Autoradiography, 6

B cell lymphocyte, 50
B cell lymphoma, 27
B23 protein, 125
Bcl-2, 168
Biotin, detection systems, 6
Biotin–avidin detection system, 62
BK 19.9 antibody, 105
Bone marrow transplantation (BMT), 68
Breast cancer 33, 79, 106, 110, 112, 114,
  140, 147
  DNA flow cytometry, 83–5
Bromodeoxyuridine (BrdUrd), 87–89,
  105–6, 138
BU31 antibody, 116
Burkitt's lymphoma, 28

C23 protein, 124–5
C_5G_{10} antibody, 114
Caenorhabditis elegans, 159, 162, 163
Campylobacter jejuni, 7
Cancer and cancer cells, 93, 97–8, 122,
  152
  interphse AgNOR distribution, 136–8
Cell cycle, 93–4
  events and stages of, 94–9
  G_0 phase, 98, 115

Index compiled by Geoffrey C. Jones